From Technical Professional To Corporate Manager

From Technical Professional To Corporate Manager

A Guide to Career Transition

David E. Dougherty

A Wiley-Interscience Publication

JOHN WILEY & SONS

New York   Chichester   Brisbane   Toronto   Singapore

Library of Congress Cataloging in Publication Data:

Dougherty, David E., 1933–
  From technical professional to corporate manager.

  Includes index.
    1. Engineers. 2. Scientists. I. Title.
TA157.D58 1984   650.1′4   84-7236
ISBN 0-471-80707-9

Printed in the United States of America

10 9 8 7 6 5 4 3 2

Dedicated
to
Beverly

Preface

This book is directed to technically oriented individuals who are pursuing a corporate career. Its primary purpose is to help creative engineers and scientists to attain their professional goals.

From Technical Professional to Corporate Manager tells you how to get bigger raises and promotions. It also shows you how to direct your career and to reach a higher level in the corporate hierachy.

You can get a promotion, provided that you are willing to work for it. In most cases, it is a matter of direction and recognition, of standing out from your peers, and of doing more things right most of the time. In many cases, career planning, goals, direction, and hard work add up to bigger raises and promotions.

This book advises you how to use your company's resources for career advancement and how to avoid common errors. It provides insight into corporate practices and introduces you to successful executives and the techniques they used in climbing the corporate ladder.

Most of the individuals you meet in the following pages have succeeded in the corporate environment; they provide a living lesson in how to work effectively, how to get ahead, and how to enjoy life. A few individuals are introduced by their first names and initials, and they too have been successful. However, they are used to illustrate practices that might be better avoided in your quest for advancement.

You also encounter a few unsuccessful individuals in the following pages. Their names have been changed to avoid embarrassment. Some readers may associate their own activities with those of the unsuccessful individuals. Those readers will see that it is not too late to change, to put past practices behind them, and to start fresh.

Selling your ideas and yourself is emphasized throughout the book. You learn to use innovation as a springboard to success and how patents and other forms of proprietary property can help you do a better job and advance your career. Actual case histories illustrate the intricacies of patents and how they can be used in selling your projects and, more importantly, yourself.

Technically trained individuals include those with a 2 year Associate of Arts degree, the professional engineer, and the scientist with a Ph.D. They are, for the most part, referred to as engineers. A large majority of engineers are members of the male sex. For this reason, the masculine form of the personal pronoun is used. No discrimination is intended.

As you read the book, recognize that each point is designed to get you a bigger raise and a promotion. For those of you who are willing to put forth the extra effort, and sincerely want to get ahead, read on!

DAVID E. DOUGHERTY

Weston, Connecticut
June 1984

Acknowledgments

I appreciate the contributions provided by the many capable engineers, executives, lawyers, managers, and scientists with whom I have worked during the past 25 years. This book is based on their approach to industry and on their way of life.

I am particularly grateful to Bruno Miccioli and Rocky Portanova, who read my unedited manuscript and made so many valuable suggestions.

D.E.D.

Contents

APPENDICES   221

From Technical Professional To Corporate Manager

1

The Dynamics
of Success

Believe in yourself. Have faith in your abilities. Without a humble but reasonable confidence in your own power you can not be successful or happy.*

*Peale, Norman Vincent—*The Power of Positive Thinking*, page l. Copyright © 1952, 1956, Prentice-Hall Inc., Englewood Cliffs, NJ 07632.

1.1 INTRODUCTION

You can succeed in a corporate environment by applying scientific principles to the management of your career.

Pay attention to your career is the first principle of success. In time, a mentor may help in this role. In the interim, apply the rule that God looks after him who looks after himself. It will be relatively easy for you to surpass many of your peers by recognizing what it takes to get ahead, by applying a few simple rules, and by paying attention to career goals. This is so, because many talented engineers deal themselves out of the competition for promotion. Some work themselves into a rut and become frustrated.

It is not always a question of how hard you work. Almost everyone will recognize poor old Sam, who labors diligently for long hours, makes a monumental contribution to the company, and seldom, if ever, gets any recognition. Sam's colleagues complain about the company's lack of appreciation and unfair policies which favor the less talented who do less work. His coworkers attribute their lack of success to corporate politics and take macho pride in doing things their own way. In other words, they don't become a member of a team; they don't work effectively.

Take drastic action, if you find yourself among the group of complainers. Make up your mind that you can succeed and associate with a more enthusiastic group. It is not necessary to work 16 hours a day. However, it is important to direct your daily efforts toward real accomplishments in order to earn bigger raises and promotions. Sam worked hard at being a good engineer. In doing so, he failed to pay attention to his own career.

As you may know, many corporations are no longer willing to invest in long term technical projects. An era of financial controls, a preoccupation with numerical analysis, a return on net assets, and a demand for rapid payback have made the role of technical development more difficult.

Some engineers are frustrated by financial analysis and are scornful of accountants. They look down their academic noses at the financial analyst's simplistic mathematics and are finessed out of a major project.

Recognize the importance of financial management as a second principle of success. You can learn to speak their language and to apply their concepts in selling your projects.

Almost everyone wants to get ahead. However, the vast majority are not willing to put forth the extra effort. You can capitalize on their reluctance by working a little harder and by being a lot smarter. You will stand out above your peers and get larger raises and promotions.

Enjoy your work is the third principle of success. If you do not enjoy what you are doing, you are in the wrong job and should quit before you become bitter.

Almost everyone in engineering is frustrated from time to time but you can overcome this problem by analyzing the situation. A discussion with a superior or a colleague may help, but never discuss your frustration with a frustrated individual. Seek advice from a positive thinker and work to overcome any frustration.

Give the customer what he wants is the fourth principle of success. All customers want quality, service, and value. However, you should not work for less than the fair wage, since customers will usually pay a premium price for quality. Provide the service that your company wants and give value for what they pay.

You can be anything that you want to be. Have you discussed your career goals with anyone else? If you're married, is your spouse a full partner? You are both investing in your career and should share in any rewards. Are you both willing to sacrifice to reach the top? Remember, two people working together are more productive than two pulling in different directions. If you are not married, review your plan with a confidant. Obtain an objective appraisal and advice.

Establish meaningful goals with a realistic schedule is the fifth principle of success. Most successful corporations have a 1 year and a 5 year plan. Go beyond 5 years in your career planning. Establish an ultimate goal. Do you want to be a vice president of R&D? At what age can you hope to reach that position? What are the steps along the way, and by what age should you make each step? Review your program at least once each year and evaluate your performance as compared to your plan.

Some large corporations have a dual ladder for advancement. One is for management and the other is for professional scientists. The roles are quite different and the pay scale is usually weighted toward management.

Nevertheless, if you don't like administration, motivational problems, and discipline, don't try to fill a management role.

It's not what you do, but how you do it, that will determine how high you can go in the corporate hierachy. Work hard and direct your efforts to maximize your probability for success.

1.2   ENTHUSIASM

Enthusiasm is like a whirlwind. It is a dynamic quality. And it can be acquired. You can make yourself enthusiastic, you can build enthusiasm into your personality . . . .*

Ted Welton is one of the most enthusiastic individuals I know. Ted, a consultant, served as a vice president of R&D at Kennecott, group vice president of R&D at the Carborundum Co., and president of Calgon. He is enthusiastic about his projects, his work, his company, and himself. His enthusiasm is infectious.

Enthusiasm usually increases as you progress up the corporate ladder. Look at the successful executives in your company. Does enthusiasm permeate the executive ranks? If not, it may be time to look for another job.

Get involved in your work if you want to be enthusiastic. Find something exciting, a project that is challenging, whereby you can make a contribution. Recognize the importance of enthusiasm to your company and to your career.

If you think that your job is insignificant, consider why so many cars fail to finish the Indianapolis 500. In 1967, Andy Granatelli's turbine-powered car held a commanding lead, with only a short distance to go. The failure of an inexpensive part stopped it in its tracks.

Get enthusiastic about yourself. Enjoy life. Look at the bright side and help others. You personally will benefit by this approach. Would you enjoy associating with a group of complainers; would you select someone

*Girard, Joe, *How to Sell Yourself*, page 75. Copyright © 1979 by Joe Girard. Reprinted by permission of Simon & Schuster, Inc.

with a cranky disposition to motivate others? Do you think that the executives of your company would promote a complainer?

Mike Herbert is extremely bright, has an excellent education, and has many of years of experience with a fine company. Mike is almost always late and frequently dozes at his desk after lunch. He commented that he and his wife regularly watch the late movie and stay up until 2 or 3 A.M. Mike is too tired to be enthusiastic or to live up to his potential.

Be happy. There are many reasons to be happy—a good education, a good job, and an opportunity for advancement. You live in a land of opportunity and are free to change jobs. You should be enthusiastic about both your opportunities and yourself.

One summer evening, my college-aged daughter asked me for advice. Elisa had taken a summer job as director of a gymnastics program at a YMCA camp. She was concerned because her assistants were developing a case of the "blahs." Elisa told me that between classes they would lay down on the mats and would not want to get up for the next class.

I suggested that she talk with them and explain the importance of enthusiasm for motivating the children in the incoming classes. I also suggested that she be enthusiastic and find a way to keep them active between classes.

Elisa asked her assistants for help in planning and cooperation in making their program the best in the camp. She also added music to the program and had an enjoyable and enthusiastic summer.

Ted Welton taught me a lot about enthusiasm. On more than one occasion, he interrupted a hectic schedule to counsel me on career planning, even though I was not part of his organization. Ted is enthusiastic about helping others. He is also enthusiastic about the work of his subordinates and schedules meetings for breakfast, lunch, or dinner. He seeks their participation in the overall program. Ted also applies my 5% rule; that is, an extra 5% of effort separates a successful executive from a nonsupervisory engineer.

Let's refer once more to poor old Sam, the hard-working engineer who puts in long hours and saves hundreds of thousands of dollars a year for the company. His projects run smoothly and are successful, yet Sam has been passed over again and again. Sam lacks enthusiasm.

Make up your mind to be enthusiastic. It will pay off in more ways than you can imagine.

1.3 INNOVATION

You can innovate your way to success if you understand the role of innovation in advancing your career.

What is Innovation? Innovation is the introduction of something new—an idea, a method, or a device. It is like invention and requires a high degree of creativity. An invention is a solution to a problem, often a technical one; innovation is the commercially successful use of the solution.* In other words, innovation is a market oriented approach to invention. However, you do not have to make an invention to be innovative. There are many innovative approaches to marketing and innovative ways of doing business. None of which would be inventive or patentable. In reality, innovation is based on your ability to see beyond the obvious and your willingness to take a new or different approach to solving problems.

Before going farther, you should realize that there are two approaches to corporate growth and profitability. The first is internal growth, which includes increasing market share and introducing new products. The second approach is acquisition. Therefore, when you sell one of your projects, you will have to show that a potential return on an investment in your project will be greater than an investment in a new company or in an advertising campaign.

"Innovation in new products and continuing improvement of existing products is the lifeblood of AMF,"* according to Tom York, Chairman of the Board of AMF Incorporated. It is no longer sufficient to offer the same old product or to merely copy the competition. Even if you can avoid patent infringement problems, studies have shown that the first one on the market with a new product makes significantly more profit than the second.

Therefore, you can help your company to be first by being innovative or creative. A market oriented approach will also distinguish you from your peers and help you in selling your projects, in getting management's

---

*Bacon, F. R. Jr., & Butler, T. W., *Planned Innovation*, page 12.
AMF Patents—A Corporate Manual, page 1.

attention, and in getting recognition. It may result in early consideration for promotion.

Don't overlook the importance of innovation or confuse innovation with invention. Don't concentrate on the technical aspects of your job. Be creative and direct that creativity to filling a need in the marketplace, and don't forget to get recognition for your contributions.

In the late 1950's, a cartoon appeared in a technical magazine, showing a chemist in a white lab coat holding a test tube. A man in a dark suit was speaking to him. The caption read—"It's a brilliant invention doctor, I'm only sorry you're not in sales so that we could get you a raise." If you don't get recognition for what you do, you probably won't get a significant raise. Don't overlook your need for recognition at the highest possible level.

The difference between a successful engineer and a technical drone is usually salesmanship. Remember poor old Sam? Sam failed to sell himself at the early stages of his career. Sam concentrated on the technical aspects of his job. He was a great engineer, but by the time management recognized Sam's ability, he was taken for granted.

J. Tracey O'Rourke is the chief operating officer at Allen Bradley in Milwaukee, Wisconsin. In the early 1970's, Tracey was working in the R&D division of Carborundum and had taken over a plastics business. He commented on the importance of reading books on success in selling and suggested reading two or four books of this type each year. Tracey is an innovative salesman, but recognizes the importance of learning something new that will make him more successful.

Don't be intimidated by innovation. It is not, as a popular myth suggests, some inbred aptitude. Innovation is based on knowledge plus the utilization of that knowledge. In many cases, you can borrow from what someone else did in a different field. You can apply a known solution to a different problem or modify an approach reached on a trial and error basis. Once you recognize the problem, you can find a solution. Just be willing to try something different.

Edward Acheson, the founder of the Carborundum Company in Niagara Falls, tried to make synthetic diamonds and invented silicon carbide. Acheson was an innovator and built a prosperous company based on his invention. In the late 1800's, he designed and patented a furnace which is still the most efficient one for making silicon carbide.

Being an innovator is like building a muscle. At first, it is difficult, however, the more you do it, the easier it becomes. Remember that corporations thrive on innovation and promote innovative individuals.

1.4   THE DYNAMICS OF PERSEVERANCE

Nothing in the world can take the place of persistence. Talent will not. Nothing is more common than unsuccessful men with talent. Genius will not; unrewarded genius is almost a proverb. Education will not: the world is full of educational derelicts. Persistence and determination alone are omnipotent.*

Perseverance is an essential trait for anyone who wants to climb the corporate ladder. Successful men and women overcome frustration because they have perseverance. For example, some executives have missed a promotion, seen a less capable executive promoted, or been unable to sell a favorite project, yet because they have perseverance, they succeed.

If you're passed over once, it may be because you are not quite ready or that the job isn't quite right for you. If you're passed over twice, it may be time to circulate an up-to-date resume. If you're passed over for the third time, analyze the situation and find another job.

It is usually difficult to decide when to abandon your favorite project. When management decides to go in a different direction, it's useless to fight city hall. If an executive vice president shouts, "I won't spend another nickel," it's usually time to give up on the project. However, when the chief executive officer believes in your project, you should fight to make it work. Be realistic. Don't give up too early and report the facts. When management says no, however, accept the decision and go on to the next assignment.

Henry Hemenway, president of Industrial Experiences of Worchester, Massachusetts, is an avid tennis player and skier. His skiing technique exemplifies his approach to management. At the top of the lift, he looks for the lift entrance and, recognizing that the shortest distance between two points is a straight line, he attacks the hill.

*Girard, Joe, *How to Sell Yourself*, page 341. Copyright © 1979 by Joe Girard. Reprinted by permission of Simon & Schuster, Inc.

Henry likes to be challenged. He accepts difficult assignments, applies management and engineering skills, considers alternatives, and makes things happen.

In addition to his perseverance, Henry has a tremendous power of concentration and the willingness to apply his full effort to the task at hand. He has a bright and charming wife, Marge, who gives Henry the support that enables him to concentrate on his job.

Remember that you are really selling yourself and shouldn't become overcommitted to a project. Clyde is a bright engineer and a capable manager. He believed in a project and placed it ahead of his own career. After efforts to find funding for an oil shale project in the late 1960's failed, management elected to discontinue the project. Clyde said that if management didn't have sufficient confidence in him to continue, he would leave. The project was scrubbed and Clyde went to another company.

Just use your project to sell yourself. Do the best job possible and if the facts dictate, scrub the project.

Whatever you do, don't ever look like a beaten dog. No matter how disappointed because you didn't get a promotion or because your project your career. There is one other thing to avoid if you want to advance your career. Never discuss your frustrations with a frustrated individual.

Discuss your disappointment with your superior. Tell him that you are disappointed because you didn't get a promotion or because your project wasn't approved. Ask where you fell down without being defensive. Ask him to level with you, because you want to do a better job and to prepare yourself for a higher position.

If your boss won't level with you, talk it over with a winner. Personnel may help, but frequently refers you back to your superior. If all else fails, try for a transfer. Make it clear that you enjoy working in your present assignment, but need more diversity. Explain that you want a new assignment and an opportunity for promotion.

You can benefit from a transfer, particularly in those cases where your superior is in a rut. If he fails to give you exposure with higher executives or to give you a candid appraisal, it may be time to initiate a transfer to another department. If all else fails, look for another job.

However, before looking for another job, try to gain the maximum experience you can at your present company. Suggest to your boss that you have worked together for several years and have learned a lot, but would like to broaden your experience. Ask him for suggestions. If nothing comes of it and it's necessary to change jobs, leave on good terms. Never threaten to leave unless you have a new job offer, preferably in writing.

Some years ago, I was working as an engineer at Westinghouse. As an evening law school student, I was interested in a transfer to the patent organization. The local personnel manager said there were too many engineers in law school, but indicated that he would look into the possibility. Sometime later, I commented to a new boss that I was going to law school and wanted to get into patents. His eyes lit up and he said, "That's great. The patent department is so understaffed they don't have time to work on my inventions." He called the manager of patents, added an excellent recommendation, and I received an offer.

Perseverance, enthusiasm, and innovation are essential traits for a successful career. In essence,

$$E + I + P = S$$

Enthusiasm plus innovation and perseverance equal success. They also add up to more dollars.

2

The Characteristics
of Success

2.1   KNOWLEDGE

You can use your knowledge to gain diversified experience and to get ahead. You can do this because progressive companies move bright engineers from one assignment to another for diversified experience. Many of these bright engineers are placed on the "fast track" for promotion. Sometimes poor engineers are moved several times, but not for diversity. Their bosses find an opportunity to unload them.

Alex Goldstein is an example of a bright young executive. He arrived at the Carborundum Company as a new M.B.A. and a numbers oriented financial analyst. Alex was very bright, but was a bit brash. He is probably still brash, but at 29, after serving as a division controller and director of marketing, he became a division vice president and general manager. Alex was successful because he was bright, worked hard, and learned from his experience.

As a division general manager, Alex became involved in a difficult law suit. He listened carefully during his preparation as a witness and did exceptionally well under cross-examination.

In Europe, Alex made an impressive presentation to a number of government organizations. His astute judgment and willingness to assume risk opened a number of business opportunities.

He invited a group of engineers to his plant for a 2 week training program, to learn how and when to use his products. Alex wanted them to work with his people and to become familiar with his operations and his manufacturing capabilities. Alex created a demand for his products and paved the way for a major sale of sophisticated technology.

Your career can become more spectacular if you follow Alex's example. Listen before you venture an opinion. Find out the facts before you jump to a conclusion. Use your mind and your education and learn from your mistakes. Remember that corporations hire and promote bright people. Don't hide your knowledge. Use it to get ahead.

However, an unusually high IQ can be a disadvantage. Some bright people become lazy and arrogant toward those with an inferior intellect. Don't make this mistake; use your intelligence wisely and follow Alex Goldstein's example.

2.2   DIRECTION

The ability to recognize and concentrate your efforts on important matters is far more advantageous than a high IQ. For example, a racing driver learns quickly about the difference between rolling friction and sliding friction. When he locks the wheels at high speed he loses control of the car. It is equally important for you to control your progress. You can do this by directing your efforts and applying your knowledge to control your career advancement.

Harry Jones is a bright young lawyer who is known for his attention to detail. He spends hours of his time on projects of minor importance and is usually overloaded with work. Harry doesn't differentiate between the important and the unimportant jobs and has been passed over for promotion.

Successful managers understand the 80/20 rule: 80% of the profits come from 20% of the product line. Therefore, successful managers put 80% of their efforts on the 20% of the work that is most meaningful.

You can benefit by applying this rule. Just learn to prioritize your projects and to devote most of your time to the most important ones. Use other jobs to fill in when an important job is being held up. Then attack the lesser jobs, get them done quickly, and be rid of them.

It is important for you to analyze your performance regularly. Determine the status of each of your major projects. What can you do to expedite them? List your accomplishments and shortcomings and review them with your superior—part of his job is to help you.

Theoretical knowledge is almost worthless, until it is used to solve a problem. Solving problems will help you to climb the corporate ladder. This is one of many ways that you can direct your efforts and use your knowledge to become more successful.

2.3   PREPARATION IN ENGINEERING

There is no substitute for preparation. It is one of the most valuable tools in a successful executive's portfolio. Preparation makes the difference between a good lawyer and a poor lawyer. It also makes the difference between a bright, successful engineer and a technical drone.

The first rule of good salesmanship is preparation. Know your products before you try to sell them. Know yourself before you try to sell yourself. Alex Goldstein follows these guidelines to achieve success. He also recognized his limitations and worked hard to overcome them.

Have you ever noticed the difference between a good lecturer and a poor one? The difference is usually preparation. Even extemporaneous speakers owe their success to practice. Good speakers know their subject and exude confidence, which comes from preparation.

One bright lawyer (now deceased) never passed the examination to practice before the United States Patent Office. He graduated cum laude from law school, had a good technical education, and passed the state bar exam. Shortly before taking the Patent Office exam for the third time, he was studying books on tournament bridge. He thought he knew all the patent law he needed to pass, but failed again.

Ted Welton utilized his staff in preparing formal presentations. For example, Bruno Miccioli, his director of development, knew what was important to management, gathered the facts, and presented them with visual displays and numerical analysis. Early morning meetings and late night sessions resulted in polished, effective presentations.

You too can make an effective presentation. Preparation, including rehearsals, is usually required. Follow the example of Ted and Bruno, who usually had several rehearsals before making a formal presentation. Try to make two or three presentations to your wife and one to your associates before making formal presentations to management. If you accept criticism, you can improve with each rehearsal.

If you don't know the answer to a question, say so. Offer to find out and report back; don't guess or bluff. Deal with facts and base any opinion on those facts.

In summary, never try to do something without adequate preparation. If it's not worth doing right, don't do it. Your career may be at stake.

Several years ago, I was asked to make a presentation to the executive group on the cost effectiveness of my operation. Several days before the presentation, Bill Wattmore, the chief operating officer, called me to his office. Bill listed specific questions that he wanted answered. This information helped me to make a better presentation. You won't always encounter a gentleman like Bill Wattmore, however, you can prepare by asking others what types of questions they might expect.

2.4   EFFECTIVE USE OF TIME

Some engineers are notoriously inefficient. This inefficiency destroys promising careers. Nevertheless, inefficiency can be easily overcome, thus enabling you to surpass many of your peers.

Keep busy and use your time wisely is probably the best advice you will ever receive. Don't forget that you and your colleagues are members of the same team. Nevertheless, you are competing with your coworkers for a share of the annual raises and for the next promotion. You are also competing with engineers from other companies, who might convince management that they are more qualified than you. In a way, you are competing with yourself to do the best job possible.

George Michael is a manufacturing engineer who has been with the same company for over 8 years. George arrives 10 to 20 minutes late in the morning and then talks to his cohorts for another 15 to 20 minutes before doing any work. They discuss their golf game, Monday night football, or some other subject of equal importance to the company.

After completing the office amenities, George checks on production. He also has a number of cohorts in the shop who are interested in golf, Monday night football, and the like. He checks a couple of machines in production and wanders back to the office. At his desk, he scans a new *Chemical and Engineering News*, makes a stop in the lavatory, and leaves 15 to 20 minutes early for his coffee break.

George takes a long break to discuss office politics, rumors, and any promising want ads. After coffee, George may return to his desk to read or go to the model shop. At 11:45, he is ready for a 1 to 1½ hour lunch.

George does communicate with his boss about plant production, what's going on in the shop, and results from the lab. They also discuss their respective golf games.

George also takes a long afternoon coffee break and leaves work 10 minutes early to beat the crowd. He would leave earlier, but his boss's boss leaves at 4:15 and it would not be politically acceptable to leave before him.

During a typical day, George handles a number of pressing matters. He makes one or two personal phone calls, arranges a weekend golf game, participates in a football pool, discusses the weekend upsets, and wonders how soon he will be promoted. After all, Dick got promoted last month with a year's less service and, in George's mind, after spending almost as much time at coffee breaks as George.

George particularly enjoys meeting with vendors and telling them about proposed increases in production and new opportunities for their products. He is receptive to trying their products. Besides, lunch with a vendor means going to a restaurant, having a cocktail, and enjoying at least 2 hours of idle talk.

George is a bright and capable engineer, but he is in a rut. He has lost his perspective and believes that the company is paying him for what he knows. He is an extreme case and would not be tolerated in many companies. Unfortunately, many engineers are influenced by his habits.

Progressive companies promote individuals for what they do, as distinguished from what they know. Don't be a loser or become cynical. Attitude deterioration is a common malady which is infectious and highly contagious. Avoid George and his colleagues before they destroy your career.

If you don't fit the corporate mold, or are not willing to work within the corporate structure, don't expect to get ahead.

Walter K. is a bright, energetic engineer who is very productive, uses his time well, and has a number of significant accomplishments to his credit. The trouble was that Walter arrived at 10:30 to 11:00 A.M. and did his most productive work after 4:30 P.M., when he had the entire laboratory to himself.

Even though Walter was productive, his boss wanted Walter to work regular hours. He recognized Walter's accomplishments, but would not tolerate his idiosyncrasies. Walter realized that he did not belong in a corporate environment and returned to the academic world.

If your efforts are not appreciated, find out why. You may be at fault. If you do your part, and your employer does not, it may be time to find a better job.

2.5   TIME CONTROL

Successful people are almost always busy and direct their attention to business. In doing so, they are also working for their own advancement.

Don't let the inefficient use of time scuttle your career. Make the effective use of time a habit. Each day resolve to do a good job and to concentrate on the job at hand. Avoid the procrastinators. Ignore their jibes and get the most out of your job.

To use time effectively, concentrate on your most important projects. Do you remember the 80/20 rule? Spend 80% of your time on the 20% of your work that is most important to the company.

2.6   TIME FOR PLANNING

Allocate a portion of your time to planning. List your ultimate goal, interim goals, and a realistic time schedule. Career planning and projects planning should follow the same format and incorporate the same practices. Periodically, review your progress and analyze what you have done well and what needs improvement. Adjust your time schedule and rethink your plan.

You can become more effective by establishing goals. Identify goals for each project and take a systematic approach to meet those goals in a timely manner. Techniques such as critical path analysis will help you to complete your project on schedule. Don't forget to apply these same techniques to your career planning.

Don't work overtime to make up for long coffee breaks. Your reputation for taking long coffee breaks may overshadow your efforts. If you work long hours and get caught up, initiate a new project or devote time to planning. Don't read the *Wall Street Journal* at your desk. Read newspapers at home or before and after normal office hours.

2.7   TIME FOR READING

Some engineers are avid readers who read for self-improvement and relaxation. Reading may improve your mind but should be done in moderation.

Be efficient and allocate time for good books. Books on sales, positive thinking, business, and culture can expand your horizons. They can also help you to express yourself and to further your career.

Discipline yourself and avoid reading the same thing several times. If you read *Newsweek, Business Week*, and the *Wall Street Journal*, about 80% of the material in each publications is identical.

Newspapers and magazines usually put the most important information in the early part of an article. Skim them rapidly for the essence of the essay. Read selectively and limit yourself to those articles or portions of articles in which you have a genuine interest.

Try the following experiment. Arrive at work 30 minutes early and read *The New York Times* or *Wall Street Journal*. Did you use the entire half hour? Repeat the experiment but force yourself to read the paper in 10 minutes. You will cover the same material and have 20 minutes for productive work. You can use the extra 20 minutes to get ahead.

Reading can be a powerful stimulus to your career. Read at least one self-improvement book per quarter or books by or about successful people. Look for clues and motivation that can help you as you climb the ladder to success.

2.8   TIME AND TELEPHONE

You can use a telephone intelligently to advance your career. At times, your brief conversation can be as effective as a personal meeting, and far more efficient. At other times, your short conversation will be more efficient than lengthy correspondence. Nevertheless, your use of a telephone needs control.

Don't use the telephone to discuss Monday night football, golf, or the like. If a 5 minute phone call extends to 20 minutes you have lost 15 minutes. If you lose 15 minutes per week on nonbusiness conversation, you will lose over 12 hours a year, 15 minutes on the telephone each business day wastes almost 8 days per year.

By contrast, "Good morning John, what can we do for you?" leads to a business oriented conversation. If the caller goes off on a tangent, say that you are busy and offer to get back to him later. Avoid conversations with losers. Don't let them drag your career down with theirs.

A good secretary should screen your calls. Some executives have the secretary accumulate all but urgent calls and pick an otherwise unproductive time of day to return them. One lawyer followed this approach, but lost his clients because he was seldom available when needed. Don't make this mistake.

2.9   KEEPING TRACK OF TIME

Joe Cesarik, a corporate lawyer for Sohio's Industrial Products Co., keeps a meticulous record of his time.

This practice is common with lawyers in private practice, because if they fail to bill their time they don't get paid. Don't wait until the end of the day to reflect on what you did. Keep a running record. It's really a lot easier, and you avoid overlooking something important. It can do wonders for your career.

Try this experiment; start by accounting for half hour intervals of time. Decrease the increment to quarter hours and then try to account for 10 minute intervals. In most cases, productivity goes up exponentially as the increment is decreased to about 10 minutes.

For example, if you spend 2 hours and 40 minutes designing an experiment, write it down and go on to the next project. If you wait until the end of the day, you might record only 2 hours and wonder what happened to 40 minutes.

Joe Cesarik is successful because he works hard and makes good use of his time. If you follow Joe Cesarik's example, you can be more successful and proud of your day's accomplishments.

2.10   MAKING TIME FOR AGGRAVATING TASKS

You can overcome a tendency to put off those nuisance jobs by breaking them down into manageable portions.

Karl W. Brownell, a patent attorney, taught me another approach to managing lower priority jobs. Karl separates his work into high priority and low priority piles. Periodically, he takes a morning to clean up the entire batch of low priority items. It usually takes less time than anticipated and leaves you with a sense of exhilaration.

2.11   PLANNING FOR EFFICIENT USE OF TIME

Planning is an important executive trait. Use it to gain control of your time and to work more effectively. If you find that your most productive time is early morning, devote that time to your most important or difficult projects. Schedule meetings for the less productive times.

One engineering manager at Westinghouse was perplexed when a project review took an entire morning. He scheduled the next one for 11:30 A.M. and completed it in 45 minutes.

If you really want to move ahead, get a head start on your competitors. Arrive a half hour early and organize your work. Start without interruptions and be off and running while your peers are trying to get organized. Thirty minutes for planning and organization will almost guarantee a successful day. List your most important job and get your adrenalin flowing. Besides, it's usually easier to arrive 30 minutes early, since you will avoid heavy traffic.

In summary, you only have a limited amount of time; use it wisely and advance your career. Wasted time leads to a wasted career.

3

More Characteristics
of Success

3.1   YOUR ATTITUDE

You can accomplish whatever you wish with a proper attitude. Develop a take charge attitude, make something happen, and build confidence in yourself. "Selling yourself successfully depends a great deal upon your attitude toward others" and "on your attitude toward yourself,"* according to Joe Girard, America's greatest salesman.

If your top executives don't have confidence in themselves, look for another job. If you don't have confidence in yourself, get help.

Alex Goldstein (introduced in the previous chapter) has a positive attitude. He knows that some of the things he tries will not succeed, but is confident that he will have a good batting average. One winter Alex, and I spent a week in Moscow. It was cold and overcast, yet we enjoyed the business meetings and a brief glimpse at the beauty of the Russian city. A year earlier, I spent a week in the same city with an individual who complained about the weather. The complainer also overlooked a fascinating culture. Alex made a far better impression on the Soviets because of a positive attitude.

Young engineers may feel intimidated by a senior executive. Recognize that the senior executive is asking for your opinion. He is also interested in facts that only you can provide; an executive is a generalist and is probably a bit rusty in his knowledge of technology.

Beware of attitude deterioration. Don't harbor frustration. Don't listen to unsuccessful engineers or let them influence your attitude. If you are passed over for promotion, disappointed in a raise, or frustrated by something at work, bring it out into the open. Confront your frustration, overcome it, and move on to a promotion.

You can improve your attitude with minimal effort. Take a personal inventory: you have a good education, a job, enough to eat, and a place to sleep. Many of you have loving families, good friends, and sufficient money to provide more than the essentials.

Take pride in your accomplishments and generate enthusiasm for your endeavors. You have the ability to do a good job and an opportunity to contribute to society. You may be surprised by some of your op-

---

portunities. For example, in 1973, I was a delegate to the USSR to exchange ideas on patent management and technology transfer. I still feel a tremendous sense of accomplishment in helping two great countries to better understand each other.

Insecurity leads to a poor attitude. Everyone makes mistakes but you can profit from yours. Almost every successful executive makes mistakes; the difference is that they have learned to accept risk.

Edward H., a salesman with a bad cold, did not feel like going to work one morning. He was behind in his quota, so he took two of his wife's prenatal vitamins and resolved to do the job. He closed more sales that day than he had ever closed in a month.

Ed discovered the importance of attitude. He analyzed his success and realized that a certain percentage of potential customers would buy his product. He viewed each customer as a challenge and as an opportunity to excel in his job.

Ed says, "If I don't think a customer will buy, I can't sell." He does not suggest prejudging a potential customer, but emphasizes the importance of attitude in doing his job.

Make certain that you are sold on yourself. As an engineer, you have completed a very demanding educational program. You have learned to analyze problems and apply scientific principles to find solutions. When you are confronted with a tough problem, be thankful, because you wouldn't be needed if it was easy.

You sold yourself to get your present position, and with a positive attitude you can sell yourself again and again.

Purge yourself of any bitterness or animosity. Bitterness is a terrible waste of energy. It is counterproductive and can destroy individuals and their careers. If someone has really alienated you, there is only one way to get even. Forget it, because every sour thought works against you. Don't clutter your mind with vindictive thoughts, don't get dragged down to their level.

Jack Pingry is a very bright engineer with an M.B.A. At one time, he really aggravated me by doing something of such minor consequence that now I can't recall what it was. Nevertheless, I let it fester. One day, Jack charged into my office and demanded to know what was bugging me. I told him and his anger immediately dissipated. In a different tone, he said that he was sorry and wished that I had brought it up earlier.

Jack confronted adversity; I had tried to avoid it. We became good friends and I still appreciate the valuable lesson he taught me. If he had followed my approach, I would have maintained a poor attitude and missed an opportunity to develop a good friend.

Joe Girard, in his book *How to Sell Anything to Anybody*, develops a "rule of 250." The rule is that most people know 250 other people. Make one enemy and you have a potential for 250 enemies. Make one friend and you have a potential for 250 more friends.

Treat everyone as you would like to be treated and you will be far more successful. If you don't like someone, recognize that it could be based on a misunderstanding. Be cheerful. Develop a sunny disposition and a positive attitude.

If in doubt, look at the crowd of "losers," those who are not promoted. How many seem frustrated? How many have a negative attitude? Is this the image you want to project? If it is, you are wasting your time reading this book.

Walk with a purpose and make it obvious that you have something important to do. Try to identify the unsuccessful engineers by the way they walk. Can you see the lack of urgency in their efforts?

Getting started is more than half of the battle of improving your attitude. Rise early, take a brisk shower, and get off to a great start.

Dress with care, because people who look sharp, feel sharp. Have you noticed how much better you feel when you wear a new suit? You can gain that same feeling with a suit fresh from the cleaners. Put on a new shirt or tie and have that same charged up feeling.

Take every opportunity to get charged up about your work and to work with people who have a positive attitude. Enthusiasm is catching. The alternative is a disease that will destroy your career.

Be happy to be alive. If you enjoy life and enjoy your job, you will feel better and will go a lot farther.

Some years ago, I discovered that there is one secret for making every day a success. I discovered it as I drove to work one sunny morning in January. As I looked up at the blue sky I said, "Thank you God for such a beautiful day, help me to make the best of it." I immediately felt a surge of energy and had a great day. Try it and you will do a better job.

3.2   COMMUNICATION

Nothing is more important to your career than clear communications. Why, then, do engineering schools devote so little time to English composition? The faculty is usually preoccupied with engineering skills and fail to teach their students how to advance in a corporate environment. Learn to communicate well and you will earn a promotion.

Utilize your free time by taking a night course in business communication. Even better, take a course in free-lance writing and read *On Writing Well* by William Zinser and *How to Write Like a Pro* by Barry Tarshis.

Every report you write should help to sell you as a bright articulate individual.

Some years ago, I wrote a report for the chief executive officer. It was extremely important, carefully prepared, and reduced to 1½ pages. Before submitting it, I asked Ted Welton for his comments and was told that the CEO never read more than 1 page. He suggested that I use wide margins and highlight significant points by indenting and preceding each with a dash.

The suggested format was preferred by the CEO and resulted in a concise memo. It pays to be concise and to give the customer what he wants.

In writing longer reports, use an executive summary. Include 1 or 2 paragraphs that state the essential points. Be clear and complete.

Bernie Chanin, a partner in a large Philadelphia law firm, is a fine trial lawyer. Bernie has a vast vocabulary and an eloquence characterized by using precisely the right words on every occasion. He uses his knowledge of the English language in advancing his clients' causes.

Prepare yourself for future opportunities by learning to speak and write like a professional. Don't risk the loss of a promotion by offending someone with abusive language.

Use lay terms in management communications in place of technical jargon.

Doctor Jesefski, a dairy chemist from the University of Montana, was a witness for my client in a lawsuit involving the manufacture of sour cream.

The defendant's expert described the process in terms of exponential growth and hyperbolic functions; Dr. Jesefski used a sponge as an analogy. The judge asked Dr. Jesefski a number of questions, received clear answers, and ruled in favor of my client.

Be sensitive, listen and answer the question asked. If you don't understand a question, ask for clarification. If you don't know the answer, don't bluff. Say that you don't know, but offer to find out. Give your superiors what they want and be concise.

Use diplomacy in your communications. Karl Brownell is an aggressive but diplomatic patent lawer and a master of diplomacy. The Corporate President (who was concerned about several lawyers leaving the company) asked if the company was paying the attorneys enough. A senior vice president (Karl's boss) replied that they were. Karl recognized that pay was the problem, yet did not want to contradict his superior in front of other executives. Karl added, "what Don said is true, since we are paying enough to get young attorneys, but we are not paying enough to keep them." A few days later, Karl received authorization to hire an experienced attorney and to pay a competitive salary.

Recognize that communication is a vital part of your job. Do it well and join the group of executives who are good communicators.

3.3 MEMORY

You can use your memory effectively to advance your career. There are books written to help you improve your memory. One or more may be appropriate if you have a particular problem. In most cases, a few simple rules will enable you to use your own natural ability.

As an engineer, you are familiar with computers and can apply computer principles to improve your memory. Your mind is like a computer's memory and has a limited storage capacity.

Therefore, don't fill your personal computer with garbage. Focus on information that is important for your career. Knowing what happened on television last night is not going to help your career; yet that type of information can fill your memory bank and prevent the recall of an important fact.

Don't rely on your memory when it's not needed. Keep good written records. Tests have shown that memory decreases in proportion to time. Write it down while it's fresh. Remember that lawyers, farmers, and engineers who keep good records earn more than their counterparts who do not. Written records are a substitute for a good memory. Written records also free your memory to store and recall important facts.

Think of putting your ideas onto a floppy disk. A 3×5 card or an engineering notebook will work just as well. A few years ago, an executive recruiter telephoned and asked me for an interview. He mentioned that Monday was inconvenient, because he was taking his son to college in Ohio. At the subsequent meeting I said, "I hope you had a good trip to Ohio and helped your son to settle in college." The recruiter complimented me on my memory. I had used a 3×5 card as a memory jogger.

Jot down personal notes about individuals. If one of your associates has a son in athletics, make a note of that fact. Before meetings, review your notes and ask the associate about his son. It can help your reputation for memory and thoughtfulness.

You can also remember people's names if you try, and at the same time help your career. When you meet an individual, repeat his name. If you have to ask him to repeat it, look at him and in your mind repeat his name over and over several times. When you get back to your desk, write it down and add any other pertinent information.

A young lawyer prepared a complex contract and mispelled the client's name. A senior partner saw the error and corrected it. He advised the young lawyer that the client would not read the contract but would see his name at the time of signing. The senior partner commented that the client will assume that if you can't get his name right, you must not be very bright.

3.4 PACKAGING

Your appearance is important to you and to your employer. Twenty years ago, you could always spot an engineer. He carried a slide rule and wore scuffed shoes, white socks, and a rumpled suit. The hand calculator has replaced the slide rule, but the scuffed shoes and white socks frequently distinguish an engineer from a financial executive. W. Thomas York, Chairman of the Board at AMF, is a meticulous dresser, as is Fred Ross, Chief Executive Officer of Raymark. Both executives recognize that their image is a reflection of the company. How many top executives have you seen wearing safety shoes and white socks?

What type of image is reflected when you look in the mirror? If you look like a successful executive, you may become one.

It is not necessary to buy an expensive wardrobe. Select your clothes carefully and recognize that wool suits look better than polyester suits. Avoid sports coats and leisure suits. Dress conservatively and have your suits cleaned and pressed regularly. If you have trouble with colors, enlist the aid of a spouse. An attentive spouse should also remind you to polish your shoes and select matching ties and shirts.

If you lack the ability to match ties and suits, get help. Keep a list: give each suit a number and list the ties and shirts that can be worn with each suit. If you dress well, you will be more confident and more effective in selling your projects and yourself.

3.5 SELLING YOURSELF BY SELLING YOUR IDEAS

"The World's Number-One Product is—Me [you], and no one can sell Me [you] better than myself [yourself] . . . ," according to Joe Girard.* There is one very effective way to sell yourself. Tell your boss about an idea that will save the company hundreds of thousands of dollars each year. In essence, it is easier to sell your ideas and indirectly to sell yourself.

The unsuccessful engineers may argue that it is difficult to sell an idea. They may point out one or more great ideas which the company turned down and say that other companies have adopted those same ideas and become more successful than your employer.

Don't listen to a loser. If the grass is so much greener at another company, why hasn't he moved? The fact is that the other company may

---

*Girard, Joe, *How to Sell Yourself*, page 16. Copyright © 1979 by Joe Girard. Reprinted by permission of Simon & Schuster, Inc.

be successful, because it has eliminated the complainers from its organization.

You can become a better salesman. It takes work to sell your projects. If you don't, your peers will sell theirs and win the larger raises and promotions.

There are techniques for improving your ability as a salesman. The first step is to recognize the importance of selling in science and engineering. Read books on sales and learn how to sell yourself.

Bruno Miccioli, as a director of development, recommended that you "concentrate on one subject in making a sales presentation." He believed that, "It is more effective to make short presentations on a single subject rather than losing the audience's attention midway through a second subject." Bruno says, "concentrate on the business opportunities and dwell on the market."

Focus on what is important to the customer. Focus on the company's needs and base your ideas on filling those needs. Never try to sell your ideas without adequate preparation. Look beyond the technical merits and describe a business opportunity.

Financial analysis may be the key to selling your project. What is the projected payback on an investment in your idea? A friend in finance can help in analyzing the cost of capital and other investment criteria. Remember that financial managers frequently control the purse strings and must be sold on the value of an investment.

Your friends in marketing can also help to sell your project by showing a customer's need that will be satisfied by your idea. Ask for their views on market share and on what type of reaction they would expect from competitors.

Technical feasability may be the least important consideration in selling your idea. For example, in one high level meeting, the financial executives spent over an hour discussing the financial ratios attributable to a new business opportunity. No one asked if the idea would work, even though it was not technically feasible.

You will probably have to convince the technical management that your idea is sound, but don't overlook marketing in this presentation.

Remember that you are now, and probably always will be, your own best salesman. Besides, you will not get promoted until you sell management on your ability.

3.6   CAREER ORIENTATION

A man who is self reliant, positive, optimistic and undertakes his work with the assurance of success magnetizes his condition. He draws to himself the creative power of the universe.*

There is probably nothing as important to you as career orientation. A ship without a rudder won't reach port. An engineer without career orientation will not find job satisfaction. Job satisfaction is essential for advancement, self fulfillment, and happiness.

Make up your mind about what you really want to do. Establish goals and analyze the amount of time and effort required to reach them. Write out your goals and a realistic time schedule. List the steps required for each goal. If you want to be a company president by age 55, map out your steps, such as vice president by 50, superintendent by 40, assistant superintendent by 35, department chief by 30, and so on. Be realistic.

Sam Marcus, director of human resource development at AMF, suggests one approach for career advancement. Sam says, "Trace the careers of your company's top executives. If a majority worked at one division, it would be wise to spend time in that division." In other words, try to pattern your career after someone you admire within the company. If there is a young vice president who received rapid promotions, ask yourself what he did that was right. Can you succeed in the same time frame?

"It is your responsibility to communicate your interests to management in the early stages of your career," according to Sam Marcus. However, in setting your goals, don't underestimate experience. An early promotion could reduce your opportunity for future raises, if you lack the experience for the job. Get the experience you need for later moves.

Sam also emphasizes the importance of "telling management what you are doing and that you should not wait until you complete an MBA to seek consideration for an advanced position." Sam points out, "You have to make the system work for you."

You can also position yourself for advancement. For example, if you start on a project in research, try to move with it through the development

*Peale, Norman Vincent, *The Power of Positive Thinking*, page 118. Copyright © 1952, 1956, Prentice-Hall Inc., Englewood Cliffs, NJ 07632.

and commercialization stages. Another approach is to seek a lateral move for diversity. "After 3 to 5 years in research as an engineer or scientist you should get line experience," according to Dr. Dean Batha, director of development at Fiber Materials International.

On weekends enjoy your time with your family, get enough sleep, and think about advancing your career.

Misery at home can adversely affect your career. Howard J. accepted a position as vice president with a new company in a distant city. It represented a promotion and the opportunity he wanted. However, his wife became so upset about leaving her friends that Howard withdrew his acceptance. Since his old job had been filled, he took a lateral transfer within the company. Howard lost his opportunity for career advancement because he failed to communicate with his spouse.

By comparison, Ned Howell, a successful entrepreneur, told me how he and his wife developed a 5 year plan around his career. Once a year they spent 2 to 3 days at a resort, reviewed their progress, and established a plan for the coming year. They each listed what was important to the other over the next 1 to 5 year period. Ned and his wife had a successful partnership, avoided misunderstandings, and worked together for a common cause.

On the other hand, frustration in your job can destroy your marriage. Some executives sacrifice their marriage for career advancement. It is not necessary or worth it. If you communicate with your spouse, recognize each other's desires, and work together for a common cause, you will go farther and have a happier life.

You can also help your career by establishing a good relationship with your boss. Discuss your short and long range goals with your boss and ask for his suggestions. Get his support for professional seminars and courses at a local university. If you work hard, he may sponsor you for an executive training program.

If you decide to pursue an extra degree, have the company pay for it. Go to an accredited program and work hard for good grades. An extra degree with good grades makes you more marketable.

As you progress in a corporate career, you will encounter difficult problems. Don't become frustrated or discouraged. The next level is only a step away.

When everything else fails, there is always one way to solve any problem. I believe that it was Dr. Norman Vincent Peale who suggested the following as a way to solve problems: Repeat, "Lord, there is no problem which is so great that it cannot be resolved by you and I working together." If you repeat this prayer, He will give you the inspiration to solve your problem. Do your part to the best of your ability and He will do His.

4

The Value
of Your Ideas

*New products are the lifeblood of any firm. Without them,
opportunities for corporate growth and profitability are greatly
reduced**

*Bacon, Frank R., Jr. and Butler, Thomas W. *Planned Innovation*, page 3.

4.1　IDEAS

If you haven't promoted at least one good idea during the past year, you ought to be fired.

Many professionals have ideas, but do not do anything with them. If you make this common mistake, you will join the technical drones. It is what you do with your ideas that counts; technical drones usually go unrewarded.

The adage—build a better mousetrap and the world will beat a path to your door—is not true. Many advertising dollars are spent before a new product is accepted. A new idea must be sold to your boss and then to many others before a company will invest hundreds of thousands of dollars in a new product.

Engineers are frequently successful selling projects based on cost reductions, yet fail to sell themselves. Don't make this mistake. Don't overlook the importance of the political exposure associated with new products.

Your role as an engineer or scientist calls for the development of new products, new processes and improvements, which depend on creativity. At times, a seemingly straightforward approach may result in an invention. At other times, some inspiration will lead you to an approach that appears contrary to engineering principles. In either case, you may have made an invention.

Ted Welton, an entrepreneur, was vice president of R&D at The Carborundum Company. He was effective in selling his project because he had the ability to speak management's language. He understood marketing and could translate market share into financial performance. He convinced management to invest in his projects and obtained funding that might otherwise have gone to other organizations.

As an individual, you are competing with other scientists and engineers to obtain funding for your projects. You are also competing with members of other organizations for a share of the corporate resources. Promote your ideas effectively and earn your promotion. The alternative is to avoid recognition.

4.2　INNOVATION AND RECOGNITION

Recognition is a prerequisite for promotion and innovation is a way to get recognition.

Do you remember being interviewed for your present position? How often were innovation and creativity mentioned? Were you offered a challenging environment and an interesting opportunity? It's amazing that so many scientists and engineers, men and women with excellent educations, can be told over and over about the importance of innovation and continue to ignore it.

Do you think that you were misled or that the company failed to live up to its promises? If so, there are three alternatives. You can quit, join the technical drones, or be realistic and ask yourself the following questions. What have you initiated and failed to complete? What have you done that will benefit your employer?

Innovation is not a result of natural aptitude. Innovation is the result when you use your education to find unique solutions to problems. Don't think of your ideas as obvious. Think of your ideas as the natural result of your own creativity.

Get excited about your ideas and see how enthusiasm helps in promoting your projects. Be positive, not apologetic. You will be surprised how your career takes off when you are recognized as an enthusiastic and innovative engineer.

Start by convincing your boss about the merit of your idea. Since he didn't think of it himself, he may initially assume that it's no good. Your job is to show your boss that your idea has tremendous commercial potential. Your job is to get him enthusiastic about your idea. At times, your boss may promote the idea as one of his own or give you constructive criticism and full credit. In either case, your job is to convince management that your idea has tremendous commercial potential.

Irrespective of your boss's approach, you probably haven't done your job. Your ideas need to be sold on more than their technical merit. They should be sold on their marketability and profit potential.

In some companies, a decision to file a patent application is reviewed by a committee that may include the Vice Presidents of R&D, marketing, and patents. Other companies require approval by the head of an operating unit. The filing of a patent application is then reported to management and, in some cases, a monetary award is paid to each inventor.

Dr. Robert J. Meltzer is a creative and enthusiastic individual. He is also a prolific inventor who became a vice president at Bausch & Lomb. Bob's creativity was complemented by a market oriented approach. He sold himself by selling his ideas. He made things happen and was recognized for his innovation. Follow Bob's example and you will become more successful.

4.3 PROMOTING YOUR PROJECTS

A positive attitude, enthusiasm, and work are essential ingredients for selling your ideas.

Start by analyzing your idea. Do away with the technical jargon and define your idea in layman's terms. Don't try to impress people by technical sophistication. List the advantages of your idea and any disadvantages, together with all of its potential applications. Be realistic and optimistic in estimating the development time and cost for each application.

If you haven't made friends in the marketing organization, acquire some now. Take them to lunch at a good restaurant and pick up the check. If your company won't reimburse you for a business lunch, write it off on your income tax as an unreimbursed business expense. Think of the expense as an investment in your career.

Learn about marketing in general and discuss the applications for your idea with a marketing professional. Ask your friends in marketing for suggestions on how to promote your ideas, for an estimate of market size for each application of your idea, and about price sensitivity for each type of product. One problem frequently encountered in central R&D is that scientists are too far removed from the marketplace. Your job is to tune in to market needs and to get facts from the real world to help sell your project.

Is the size of the market sufficient to justify the proposed investment in your idea? Be realistic in your analysis. Tom, a young engineer, promoted a new process for the foundry business. He assumed that the entire industry would utilize his product. Later, he found that it only applied to 10% of the market. Tom did not include an installation cost in his initial analysis. He also learned that the cost of installing the device was greater than its price and that more than one device would be needed for each casting. The final blow came when he discovered that his calculated savings had been greatly exaggerated. Tom had used the cost of the scrap metal rather than the expense of remelting it for his initial analysis. Tom was unrealistic in promoting his project.

In selling your project, emphasize profit potential. Call attention to the strength of the patent and the technical advantages. This approach will enhance your chance for promotion and may put you on the fast track.

4.4   FINDING A MENTOR

A mentor is someone who believes in you, and is willing to stick his neck out on your behalf. Mentors are members of higher management and can help to pull you up the ladder.

Every project needs a mentor but, in the beginning, you have to promote your own projects. You have to do the spade work and give others the ammunition to sell your ideas. Doing a thorough job may attract a mentor, because executives need hard workers they can count on.

Throw a ball up in the air and it will decelerate. Send an idea up the ladder and it will follow the same principle. With each step up the ladder, the importance of an idea diminishes. On the other hand, if it catches an executive's attention and is sent down for further work, its importance increases. By sending a number of well thought out ideas up the ladder, you will attract the attention of management and may be on the way to a promising career.

Jack Mummert is a retired group vice president at AMF. He has a background in sales and is fun to work with. Jack is enthusiastic about new products and recognizes the importance of patents. As a group vice president, Jack Mummert hired bright people to work in promising fields. He recognizes that new products don't always turn out as anticipated, but has confidence in people and in his own ability to sell new products.

4.5   PATENTS

"At AMF we emphasize the protection of Inventions since they are critical in meeting the challenges of a changing market,"* according to Hans W. Lange, an executive at AMF Incorporated.

Patents can serve as a foundation for your successful career. A patent is a limited monopoly that may give your company a competitive advantage.

*AMF Patents—A Corporate Guide, page 7.

A patent will enable your company to charge a higher price and will help to obtain a more favorable return on an investment in your idea.

A patent lawyer may caution that a patent does not give you the right to make, use or, sell a product. It only gives patent owners the right to exclude others from using their inventions. The problem is that someone else may have obtained an earlier patent, which might prevent the second inventor from using his invention. A patent lawyer would refer to the first patent as a basic patent and the second as an improvement.

Genius is not a prerequisite for invention. Many inventions are the result of hard work, and some result from accident. For example, Edward Acheson was trying to make synthetic diamonds and produced silicon carbide, an important invention which formed the basis for an abrasives industry.

At times, it is so easy to make an invention that the invention may go unnoticed. If an idea is new or different and useful, it is probably patentable. If it's overlooked, the company may overlook the inventor.

An invention should not be abandoned because it may be obvious to a person of "ordinary skill in the art." Patent examiners who are employees of the U.S. Patent and Trademark Office, have a reputation for rejecting patent applications on that basis. If your idea is obvious, why hasn't someone produced a product based on it? Actually, a persistent patent lawyer can obtain a patent on almost anything. This is because a patent examiner is evaluated on the number of dispositions of applications, and it is usually easier to allow an application than to continue arguing with the attorney.

Eugene Letter, an engineer at Bausch & Lomb, was initially hesitant about submitting his idea to the company's patent department. He learned to recognize the patentability of ideas, submitted numerous disclosures, and established a reputation as a creative inventor.

Several statutory requirements should be considered regarding patentability. An invention must relate to a process, a machine, an article of manufacture, a composition of matter or an improvement, an asexually reproduced plant, or a design in order to be patentable. Almost anything that is new and useful will fit into one of these catagories. There are exceptions, however, such as a method for doing business.

It is usually very easy to submit an idea to a company's patent department, even though some companies require a superior's approval.

A few superiors refuse to approve invention submissions for a variety of reasons. Your diplomacy can usually overcome this problem.

At Westinghouse's Air Arm Division, one supervisor refused to approve ideas for submission to the patent department. One resourceful subordinate discussed his ideas with this superior and asked for his suggestions. At the end of the conversation, the employee suggested submitting the idea in both names. The superior agreed to submit the idea, but only in the name of the employee. In theory, the patent attorney would have reviewed the disclosure and determined the true inventor. Nevertheless, proper entry in your engineering notebook can protect your interests.

Some years ago, a young engineer took a different approach by recording his idea and having it witnessed, before discussing the idea with his superior. He then submitted the idea in his own name and when asked why he failed to include his boss as a coinventor, he proudly displayed the earlier record. It turned out that the idea was not patentable and he had alienated his boss for nothing. He got a poor grade for diplomacy.

4.6 WHAT PATENTS WILL DO FOR YOU

Recognize the importance of patents for your company and use them to advance your career.

From your point of view, a patent is a publication and an indication of creativity. A patent is also a basis for additional funding for your project. Using it in this manner shows business acumen, which is an essential element for promotion.

Do you remember poor old Sam, the hard worker who is always passed over? Many prolific inventors are also passed over for promotion. Some don't even get their fair share of raises. They fail to sell these inventions and to receive recognition for their contributions to the company.

AMF's Patent Department, like many other corporate departments, sends a copy of each invention disclosure to the vice president of engineering. The Patent Department also notifies management when a patent application is filed and when the patent is issued. They also consult management before filing foreign patent applications. What this means to you is that the patent department is repeatedly telling top management that you are a creative individual.

You may ask about Sam. Sam got all of this attention and failed. Maybe he didn't want a promotion, but probably turned it down inadvertently by failing to promote his ideas.

Getting a patent may not result in a promotion; however, it is one way to sell yourself. Used properly, it can lead to larger raises and promotions.

How well do you prepare for your annual performance appraisal? At your performance appraisal, point out your contributions. Emphasize the importance of your patent applications and creativity and your expectation of a larger than average raise. If you don't fight for larger raises, others will. Receiving larger raises labels you as a star performer and may position you for promotion.

If a raise is smaller than expected, discuss it with your boss. After all, his job is to help you to do better. If you work hard, make meaningful contributions, and are not adequately rewarded, it may be time to sell your services to someone else.

Patents can enhance your saleability. They show that you are a creative individual and indicate that you are business oriented. Patents also prove that your present employer thought enough of your ideas to invest in their protection. When you obtain a patent, you have done more than publish, since you have provided a competitive advantage for your employer. Numerous patents may label you as a leading expert in your field.

4.7 HOW TO MAKE AN INVENTION

Professor James Bright said that on the average, it takes over 25 years for pioneer inventions to reach commercial success. With today's high interest rates, few companies can afford to invest in pioneer inventions. Hence, you should concentrate your efforts on making improvements to existing products that will reach the marketplace within a few years.

In many cases, it will be easy for you to make an invention; it is only necessary to recognize a useful idea and to call that idea to the attention of a patent attorney.

Robert Meltzer submitted an idea that was anticipated by an earlier patent. He exclaimed, "that proves it was a good idea since someone got a patent on it." If this happens to you, ask yourself why the patented idea wasn't commercialized. Can you improve on your original idea to make it a commercial success? If the earlier patent has not expired, can you avoid the claims?

Become patent oriented and analyze your ideas in terms of their patentability. Don't hesitate to ask your patent attorney for assistance. However, recognize that without commercial potential, a patent won't help in selling an idea.

In working with your patent attorney, start by defining your project and by asking for a state of the art search. This will enable you to learn what others have done in your area. For example, a state of the art search on granular activated carbon produced over 350 unexpired U.S. patents. A similar search on bras and girdles produced 750 and 450 unexpired patents, respectively. You may want to define your field more narrowly to limit the number of patents. For instance, if your field is strip chart recorder pens, limit the search to recorder pens, rather than including all writing instruments.

If you have difficulty understanding the differences among a number of patents, ask your patent attorney for help. He can show you how to distinguish a patentable concept as well as several shortcuts that will help you to review patents more rapidly. At times, you can borrow features from several patents and come up with a patentable combination.

If one approach suggested in a patent appears promising, ask yourself why the patented idea has not become a commercial success. Try to modify the idea. Chances are that you can improve on it, and in the process, make an invention.

Michael Herbert, the patent lawyer discussed in Chapter 1, is basically lazy and is characteristic of a small percentage of negatively oriented lawyers. They enjoy showing enginers that an idea is old. By doing so, they avoid the work of preparing a patent application. Nevertheless, diplomacy in selling your idea can overcome their resistance. Try to sell your attorney on the commercial importance of your idea and ask if he can find a way to get a patent on the idea.

Always try to convince an attorney that your idea has commercial potential. Be enthusiastic about your ideas, so that the lawyer will be enthusiastic in selling the Patent Office on the merits of your invention. An attorney's attitude can help you to get a patent. It can also help you in selling your ideas and yourself.

As you become more familiar with patents, you will discover that you are making more inventions. Send those ideas to the patent department and get the patent lawyers on your team. Use their help to establish your reputation for creativity.

4.8   SELLING A PATENTABLE IDEA

Don't wait for your patent to issue before selling the merits of your idea. The issuance of your patent may take several years, but this gives you time to complete the development work on a commercial product. During this period, you can also convince management that your patent (when it issues) will give them a competitive advantage.

Ted Welton involved the patent department in the early stages of his programs. Counsel worked diligently to get the best protection and were proud of the patent position they developed. When top management questioned the scope of protection, counsel emphasized the importance of key features and stated that broad coverage was contemplated.

At one stage in the development of one of Ted's major projects, the board of directors asked for an objective review of the patent position. An outside expert confirmed counsel's opinion and the board authorized building a new plant.

4.9   SELLING A PROMOTION

If you tell everyone that you are a great engineer, few people will listen, and even fewer will believe it. You can sell yourself more effectively by selling a solution to their problem. Poor old Sam took this approach, but didn't look out for his own interests. He overlooked the importance of a carefully prepared presentation and failed to use a market oriented approach.

J. Tracy O'Rourke, president of Allan Bradley, is a master at selling himself. Tracy learned the language of top management early in his career and applied marketing concepts to selling his projects. At Carborundum, Tracy scheduled frequent meetings with top management and kept them fully advised about his progress. Remember that career direction and recognition are your responsibility. Don't be bashful about your goals. Getting promoted involves more than the sale of your ideas. It is your job to position yourself in the minds of your superiors.

Attending technical conferences is also important for career advancement. Don't ask to go. Sell the concept. Say, for example, "John, it's important for me to attend the IEEE Conference in view of the two papers on corona discharge. If you don't object, I'll plan to attend." Sell the concept · early, before somone else is selected to represent the department.

You can learn about what is going on in other companies by attending technical conferences. These conferences will also help to establish you as an expert in your field. This is particularly true if you are appointed to a committee. However, you should use discretion and attend no more than one or two siminars a year. Too much travel can hurt your career.

Some companies refuse to send employees to technical conferences. If necessary, go to one or two conferences a year at your own expense. Consult a tax lawyer about deducting it as an unreimbursed business expense or as an expense incurred in looking for a new job.

At technical conferences, use your head before you open your month. You attend conferences to learn, not to tell others about your company's latest developments. Drink with moderation; keep your ears open and your mouth shut. A few drinks and an inadvertent disclosure could cost you your job and your future.

"Presenting a technical paper with company approval is an excellent way to meet people and establish a good reputation," according to Dean Batha, Director of New Products at Fiber Materials International. That reputation can be further enhanced upon your return. Dr. Batha suggests that you "review the highlights of the conference with members of your staff and forward copies of papers to others within the company."

Another way to advance your career is to take advantage of a corporation's liberal policy on tuition reimbursement. You should get as much education as possible. An advanced degree in engineering or management is far more saleable than a group of courses or a second bachelor's degree. An advanced degree demonstrates perseverance, direction, and expertise, all of which are saleable commodities.

Some companies hire managers from other companies, pay a high salary for an unknown, and overlook the talent they already have. If this happens to you, you could be at fault. One way to avoid this problem is to let people know that you are working on an advanced degree and are seeking broader responsibility.

Working on an advanced degree also gives you an opportunity to use your personnel department. Personnel recruited you and should have an interest in your advancement. When you succeed, it will prove that they did a good job. Solicit their help in career planning and their guidance in selecting courses in leadership and human relations.

If you don't have an M.B.A., sell your company on one of the executive programs. Such programs range from a Friday–Saturday program at the University of Rochester, to a 13-week program on advanced management at Harvard, and a 1 year Sloan Fellowship at M.I.T. These executive programs are usually reserved for promising individuals who are slated for promotion. These programs can do wonders for your career.

Don't overlook travel as a way to enhance your career and to give you a broader perspective. Your visits to suppliers and customers will disclose their capabilities and needs. However, excessive travel should be avoided. If you are away too often, you may be overlooked for a promotion.

4.10 JUDGMENT

Judgment is probably the most important attribute of a good executive. Men like Merlin Nelson, vice chairman at AMF, have good judgment that is based on years of experience. Some people say that judgment is difficult to learn. However, you can learn about judgment by working with top management and observing their analysis of a problem. In time, you will learn to develop your own judgment, based to a large degree on your own mistakes.

You can use perception to develop judgment. For example, observe the degree of risk that top managers are willing to assume. You can also use good communications to avoid a catastrophic result from poor judgment. Tell management what you are planning. Analyze the alternatives and risks and review your proposed action with a superior.

When you are making a presentation, be sensitive to questions and criticism. Learn what is acceptable and what can be sold by trying. Then use these abilities to sell yourself and get promoted.

5

Evaluating
Your Employer

5.1   INTRODUCTION

In 1965, I joined the Battelle Development Co. (BDC), a wholly owned subsidiary of Battelle Memorial Institute in Colombus, Ohio. Battelle was a fine organization of very dedicated people. A majority of those dedicated people were engineers and scientists.

Before accepting the position with BDC, I was interviewed by several people on two occasions. I asked a lot of questions and was favorably impressed. After all, Battelle Memorial Institute was recognized for its leadership in technology and had developed xerography and licensed it to the Haloid Company (now the Xerox Corporation).

There was little doubt in my mind that Battelle was the right organization for me. However, my one concern was their "not for profit" status. John Gray (vice president and general manager of BDC) assured me that BDC was profit oriented.

At Battelle, I learned a lot and made many fine friends. I also learned that many of the people at Battelle were marching to a different drummer. These dedicated engineers and scientists were not profit oriented.

At first, I had difficulty understanding their reasoning. For example, if I could license their technology and earn $150,000, it was pure profit. And yet, the scientist would prefer to sell a $100,000 research program, which at that time would keep about two scientists busy for a year. They pointed out that $150,000 in profit would be spent for further research, but not necessarily in their group. Besides, the $100,000 would keep their two engineers on the leading edge of their exciting technology, whereas my licensing deal would transfer the technology leadership to the licensee. In fact, my license could preclude others from doing further work in the area.

With hindsight, I made the right choice in going to work for BDC. I was also very lucky, because things worked out very well even though I had misunderstood one of Battelle's basic principles.

Actually, I thought that I had done a thorough job evaluating Battelle as a prospective employer. The problem was that I was profit motivated and, after several years, realized that the scientists and engineers were not. It seemed strange, but these people really were dedicated to doing research for the betterment of mankind.

Hence, you should carefully evaluate a prospective employer and continue that evaluation even after you have started to work for them.

Karl Brownell, my predecessor and colleague at The Carborundum Company in Niagara Falls, New York taught me many valuable lessons, one of which was that companies continuously change. Karl said, "I've worked for Carborundum for over 40 years and yet it sure isn't the same company that it was when I started." Karl also pointed out how the company had changed. Within a few years of joining Carborundum, I realized that the company had already gone through a considerable change.

Most companies undergo a continuous evolution. Some changes are dramatic, often brought about by a new chief executive. Other changes are more subtle and may be almost imperceptible. Others are thought to be temporary, such as a cutback during a recession, yet may indicate a trend that will continue into the future.

Part of your job is to evaluate and reevaluate your employer. Good companies evaluate and reevaluate their human resources, that is their employees. Doesn't it make just as much sense to continuously evaluate your employer?

Jeff Irving, a corporate psychologist, told me that individuals work better when they enjoy their job and their surroundings and when they are proud of their organization. Jeff developed a very effective in-depth interview that took several hours. At the end of the interview, Jeff could be pretty confident about whether the individual would fit in.

Today, a large number of companies rely on psychologists to determine if an individual will fit into their organization or if an individual is suited for a promotion. Sometimes, these in-depth studies are reserved for positions in higher management, and you will have to take matters into your own hands and make certain that you and your company are compatible.

Consider for a moment, how you can effectively evaluate a prospective or present employer. Actually, evaluating a company is very simple. You begin by asking a lot of people a lot of questions. You will ask a number of people about different aspects of the company. Some answers will probably conflict with others, but if you ask enough questions, a pattern will emerge.

In this area, there are not really any right or wrong answers. You are only trying to judge whether or not you will be comfortable working in a particular environment. In addition, ask yourself, "Can I work effectively and can I make a meaningful contribution to this organization?" Appendix M lists a number of questions that should be helpful in this area.

5.2   CONVENTIONAL TECHNIQUES

Have you read your company's annual report and their 10K, which is filed with the Securities and Exchange Commission? A copy of the 10K is generally available to stockholders, employees, and prospective employees. Have you gone to the library and searched for articles about your company? Have you taken the time to review technical articles published by the company's scientists and engineers?

In many cases, the personnel or public affairs department will provide copies of this information. When you consider a new job, be certain to ask for copies of any articles or interviews by the company's officers. Is the company profitable? How much has the company said about its research and engineering? How much emphasis do they place on new products? Are their new products minor improvements, or are they in fact the technological leaders in the field? Read as much about the company as possible and analyze the results.

Don't overlook financial information. I am not suggesting that you analyze the financial figures in the annual report, but that you at least ask a stockbroker for any information he has about the company. A financially healthy company can invest more in research and development than one that is losing money. As a general rule, avoid companies that earn less than their dividend for a protracted period of time.

Now that you have read the published information about your company, you can begin to ask intelligent questions. However, before doing so, exercise some caution in selecting individuals for questioning. You should also be prepared to compensate for certain biases.

When you ask a question, look the person in the eye. If you ask a clear question and get an evasive answer, be wary. Ask another question and try for a direct answer. For example, suppose you ask someone in personnel if the company uses the Hay System or a similar system for establishing salary ranges. Then ask for the company's philosophy with respect to raises. A good answer is, "We give above average pay for above average performance." However, if a company's policy is to pay below average wages, beware. Questions like these should be tempered by a statement like, "I believe in working hard and doing more than my share and would like to know if the company appreciates this kind of effort?" You can then ask, "What do they consider an above average raise in the present economy, 5, 8, 10, or 15%?" Ask if there are any limitations on raises presently in force.

The most important attribute for any company is its integrity. The integrity of the company, officers, managers, and employees is important, yet it is often difficult to get anything but an assurance to a direct question on this subject. Nevertheless, you should try. For example, if a company constantly uses misleading advertising, you should be concerned. If a company's executives are charged with using insider information or if the chief executive's children purchase a large amount of stock several days before he announces a proposed takeover, it should tell you something. I recall that the chief executive of one company bragged about how he used sharp practices to extract unconscionable deals. Several of his employees confirmed that he did not treat his employees any better.

Other companies, like The Western Electric Co., have a reputation for taking care of their employees and for high ethical standards. By way of contrast, some of the aerospace companies lay off thousands of individuals at the loss of a contract.

You should also be concerned about a company's commitment to leadership. Does your company introduce new products or merely copy those of it's competitors? Don't be misled, there is nothing wrong with copying, so long as a competitor doesn't have a valid patent. Actually, it can be very exciting to try to catch up with a competitor's new product, but will you be happy trying to duplicate someone else's new product, rather than working to develop something new?

A second important factor is a company's commitment to quality. Companies that produce low quality products usually earn a smaller profit and are not the best long term investment for an engineer or scientist. Ask engineering personnel about the company's commitment to quality and about quality control programs. Talk to distributors or retail persons and ask about their experience with a company's products. How well a company services its products is another indication of its commitment.

According to Henry Hemenway, it is even more important to understand your industry. For example, what are the service costs as compared to the installation costs? Does your company provide long term service? Ask other engineers or engineering managers about industrial practices in their field and how this company compares with its leading competitors in this area.

Ask some of the other engineers if this is a good place to work. Then watch closely for an evasive answer. Even before starting to work, observe the average age of the engineering staff. Ask yourself if there is a good cross section of experience. Can you continue to learn on the job, and can you make a contribution?

There is one good indicator that will tell you how a company treats its employees. High employee turnover is a clear indication of poor practices. Recognize that it costs the company a lot of money to hire and train new employees. However, there are times when a high turnover can be attributed to new management and an effort to weed out the less than satisfactory workers. Nevertheless, if good people leave after 2 or 3 years, there is probably a reason. Besides asking about turnover, you can ask how long several individuals have been with the company. You can also ask to talk to one of the newer employees.

Sometimes you will be lucky and will find a dynamic management and a real growth opportunity. If you are a dynamic individual, it could be a great opportunity.

Learn all that you can about your company's program for career development. In general, a company that is concerned about developing its employees is a good place to work. Pay attention to notices and courses offered. Don't forget to ask yourself which courses will help you to do a better job.

Look at the career paths of the companies executives. Have they been promoted from within or hired from outside? How many have technical backgrounds? Have they worked in operations and been moved about within the company? If most of the executives have been hired from the outside or if most general managers are former controllers, you might want to reconsider the career potential for an engineer or scientist.

In my experience, good people move to the top. Even companies that are not technically oriented tend to recognize outstanding performance. Nevertheless, if you are a dedicated engineer or scientist, you may not be happy with a company that grows by acquisition instead of from within.

At times, even experienced engineers may misread their company's philosophy. For example, the president of one major corporation was concerned about profitability and promoted an English accountant from group vice president to executive vice president. Many of the engineers thought that the company was no longer interested in research and development. Actually, the new executive vice president was appalled by some of the past practices and insisted on well thought out programs. All that he wanted was a market oriented approach to engineering and a potential return on an investment in research and development. Thus you should not jump to conclusions and should be skeptical of individuals who are frustrated in their present position.

In general, develop a good feel for your company's philosophy. Understand its goals and how it intends to meet these goals. Learn how technology fits into its plans and what you can do to help the company meet its goals. By understanding your company, you will become a better and more effective employee.

5.3   UNCONVENTIONAL TECHNIQUES

The unconventional techniques for evaluating your employer are not exotic. Actually, these techniques are similar to the conventional ones, except that they are utilized less frequently.

Your goal is to more fully understand your employer and to use this knowledge to do a better job. To this end, talk to the company's employees in other areas, such as marketing and finance. Listen to their views and compare them with those from engineering. In some companies, this is referred to as "management by wandering around," a phrase that is misleading. You don't wander around. You do go to other departments to learn. In doing so, look for answers to specific questions.

My strong dislike for the concept of "management by wandering around" stems from the image of a small number of employees who wander about slowly as though they have nothing to do. They ask dumb questions and seem more of a distraction than a member of a dynamic team.

Now consider some of the less obvious techniques for evaluating your employer. Have you talked to a representative of one of your company's major competitors? Don't think in terms of a job interview, but rather a casual conversation at a professional meeting. Ask the representative if he knew any of your company's executives when they worked with his company, or if he served on the same committee of a trade association with an executive from your company.

Talk to one of your competitor's distributors, compliment them on their products, explain who you are, and ask how they view your company as a competitor. For example, in the abrasive industry, Carborundum, Norton, and 3M were all well-known and well-respected. Each company had its strengths and its weaknesses and both were well-known within the industry.

This type of information is only a small piece in a large puzzle, yet it will help you to evaluate your company and to understand how you can do a better job.

When did you last talk to a professional headhunter (executive recruiter)? Headhunters can be a tremendous help when you are seeking another job but it is worthwhile getting to know one or two, even though you are not presently looking for a job.

A good headhunter learns as much as possible about companies within his area or within an industry. Your company may be a client of such a recruiter or the recruiter may use it as a source for candidates. A few may even try to do both, but not very often. Headhunters glean as much information about companies as possible, so that they can place or pirate an employee.

In general, the headhunters know a company and many of its executives. They are also frank in discussing this type of information. By talking to an executive recruiter, you may be surprised to learn a lot about your company.

Ask the recuiter if he has placed engineers with your company and if they have stayed or moved on. Ask him for reasons given by departing employees.

One productive approach is to contact one or two former employees. For example, as a prospective employee, it is not inappropriate to ask for the name of one or two individuals who reported to your prospective supervisor. After all, the company will usually ask you for references and will contact them. Some of the more astute companies ask the reference for the name of someone else who is familiar with your work and then contact that individual. By following the same procedure, you may avoid a costly mistake. At the very least, you should gather useful knowledge about a prospective employer.

5.4 INTERPRETING MANAGEMENT STYLE

The Carborundum Co. in the early to middle 1970's was a fun place to work. It was fun because of its management style. The Carborundum management (before the merger with Kennecott) was aggressive. In other words, they were willing to assume reasonable risks and at the same time were committed to the corporate mandate—no dips in earnings.

How would you characterize your company? One company that I recall was so conservative, it could not operate in the commercial world. They abandoned their plans for commercial ventures and concentrated on their business as a supplier for the Department of Defense. They still maintained an engineering department, but limited the department activities to producing minor improvements in their present products, or to analyzing and preparing proposals for more government business.

A number of leading technologically oriented companies rejected Chester Carlson's invention of an electrostatic copier. Then the Haloid Company of Rochester, New York, invested millions of dollars in an unproven product and became the Xerox Corporation. Many companies would like to duplicate Xerox's success, but how many are willing to invest in the early development of a new idea? How many are willing to bet their company on a new idea?

One way to evaluate your company's aversion to risk is to look at their approach to patent litigation. Ask how many times the company has been sued for patent infringement. Get copies of the cases and read them. Ask your patent lawyer about the testimony of your company's executives and for other information about the company's philosophy toward litigation.

One senior executive told me that a former employer was so conservative, they wouldn't even call one of their patents to the attention of a competitor. If your company is ultraconservative, you may have to redesign a product to avoid any possibility of infringing someone else's patent. On the other hand, an aggressive company may consider the patent invalid and assume the risk of litigation.

In general, an ultraconservative company that is unwilling to risk litigation may be unwilling to invest in a long range research program. In fact, if the company is unwilling to assert its patents, does it make any sense to invest in patents or, for that matter, in research and development?

You will also want to find out about your company's attitude toward copying. For example, one engineering supervisor thought that any engineer that copied the work of another company ought to be fired. This seems like an extreme approach, since there is no need to reinvent the wheel. And at times, it is important to reach the market rapidly. Find out what your company's philosophy is.

In general, as managements become more financially oriented, they become more conservative in their approaches to research and development. The problem is that many research and development projects are not really susceptible to financial analyses. In other words, the probability of success, time for development, and even market potential cannot be predicted with a high degree of accuracy. As the length of the project increases, the accuracy decreases. Then, when a financially oriented executive calls for a 1 or 2 year payback, he is in effect excluding long range developments and limiting his company to minor improvements in their present product line.

Also, consider your company's approach to centralized versus decentralized management. The former is frequently indicative of an autocratic approach, where most of the decisions are made at the top. In some companies, operating units are autonomous and the management style may very greatly from one unit to another.

Some engineers make a mistake by assuming that a higher degree of delegated authority relieves them of their responsibility to keep superiors informed about progress on a project. Don't make this mistake. Take every opportunity to apprise your superiors, and if possible their superiors, about your progress.

Find out how your company's executives plan for growth. Is your company leaning toward more acquisitions or is management emphasizing internal growth? In the second case, engineering will become more important and will occupy a place in the executive limelight. In general, you should be able to get this type of information by asking your prospective superior.

One key question will help you to determine the role of your organization within the overall management of the company. Ask your supervisor to define the department's mission. If he does not understand the department's role, you should be concerned about the direction or leadership provided by the company's top management. It may be that they do not consider research and development as being very important.

5.5   COMPETITIVE HISTORY

Those who do not read history are doomed to relive it. In other words, you can learn by the mistakes of others.

In general, when an individual is highly successful, he will try to repeat that successful approach in the future. Similarly, when a company launches a marketing disaster, they are somewhat reluctant to reenter that field. For example, I tried to sell an advanced welding system to a family controlled welding company. Their vice president of marketing rejected the idea, because some years earlier they had been unsuccessful in selling a welding system. He had made up his mind that they would only sell components. I wondered how long it would be before their vice president of engineering found a more challenging job.

On the other hand, Charles R. Landback (retired president of the Carborundum Company) is my idea of an exceptional executive. Chuck had been involved in two licensing deals that lost money. Nevertheless, when a new licensing opportunity was presented, he considered the opportunity on its own merits and elected to invest in the new technology.

You can learn a lot about your company by spending one or two lunch hours with a long term or recently retired employee. A discussion of the company history may uncover a recurring interest in manufacturing synthetic diamonds, a commitment to superhard materials, or a commitment to be the world leader in silicon carbide or alumina zirconia abrasives. Knowledge of this type can help you to focus your efforts on areas that top management considers important.

When you discuss your company's competitive history, find out if the company is a price leader or follower, a technical leader, and a technically, financially, or market oriented company.

Also consider any recent changes in company management and how those changes might effect your role. Be particularly concerned about the effects of a major acquisition or merger. If your company is acquired, you can be almost certain that there will be a major change. Investigate the new owner and be prepared to move.

If I can borrow a concept from Robert Townsend, author of *Up the Organization*, I would like my epitaph to read, "He never worked for a company like Kennecott." It is true that I continued to work for Carborundum for almost 2 years after the merger with Kennecott, but like so many others was fighting to keep the Kennecott management or lack thereof from destroying a fine company.

5.6   OUTSIDE ACTIVITIES AND SOCIAL RESPONSIBILITY

Ask personnel about your company's commitment to social responsibility and whether or not the executives are engaged in community activities. If a majority of the executives in your company are involved, the company must place a lot of emphasis on participation. If none of your executives are involved and relatively obscure individuals are assigned to projects like the United Fund, it is obvious that the company is not overly concerned about community relations. Some companies encourage their executives is involved. However, if your company emphasizes social company takes this approach, you may find the key to broader exposure and recognition in such activities.

Don't join in some community activity because one of your company's executives is involved. However, if your company emphasizes on social awareness, you might help yourself by becoming more active in community activities. Whatever else you do, don't overdo your outside activities. If outside activities have an adverse effect on your job performance, abandon them. Generally, you can make valuable contacts and good friends through your involvement in the community.

5.7 THE BOARD OF DIRECTORS

As a professional in a corporate environment, you ought to know the names of the corporate officers and the members of the board of directors. You should also know a little about their background.

After all, if the president of Bell Laboratories is named to your board, you can be assured of an interest in research and development. On the other hand, the complete lack of a scientist may tell you something about the corporation's interest in technology.

You might also question the objectivity of a board of directors that includes a large number of corporate officers, since the board represents the shareholders or owners and is not supposed to manage the company. Actually, the only justification for an officer dominated board is if the officers own the company.

5.8 PERSONNEL POLICY

There is probably nothing so important as a company's commitment to treat its employees fairly and to reward those individuals who make a real contribution to the company.

To begin, ask yourself if your company pays a competitive starting salary. However, don't be misled by surveys that show very high starting salaries, since they may not account for geographic differences. For instance, the cost of living in New York or Los Angeles will be far greater than living in a small town in the midwest. One of my good friends made this mistake. Bill took a job in Baltimore with a major company and received a large raise. When he moved his family from a small town in Alabama, he discovered he could no longer afford many of the things that he took for granted in Alabama. He changed jobs again and received a slightly higher salary in a small town in Pennsylvania.

Some companies pay competitive starting salaries, but for some reason fall behind by giving raises that do not adequately reflect inflation. If this is the case with your company, recognize the value of the experience in your present position and keep your resume up to date.

A more important indicator to watch is a company's propensity to promote from within. Find out if your company has a meaningful approach to human resource development and if they promote from within. If they

.don't, you should be prepared to move to a better company. If they fail to develop their present employees and offer less than competitive salaries, you should seriously look for another job.

5.9   COMMITMENT TO RESEARCH AND DEVELOPMENT

There is one key question to determine your company's commitment to research and development. Simply ask for the percentage of sales dollars devoted to research and development. Then compare that figure with the expenditures by your competitors.

However, you should proceed with caution and recognize that these figures are only a rough approximation. Look a little closer and see how much goes for real research and development versus quality control, pollution control, manufacturing engineering (troubleshooting), and the like. Then compare the figures with those for your competitors or the industry.

5.10   CONCLUSION

In summary, you should learn as much about your employer or prospective employer as you can, within reasonable limits. Much of the information is available in published reports or from a long time employee. For example, how would you characterize your company? Is it

- A high technology company?
- A consumer goods company?
- A heavy industrial company?
- A service company?
- A resource based company (oil, gas, or metals)?

Once you characterize your company, find out where it stands with its competitors and determine if you will be happy in this environment. Look for professionalism in management and for the type of people with whom you want to associate.

6

Utilizing
Hidden Resources

6.1   SURPASSING YOUR PEERS

One of your goals may be a promotion. As you climb the corporate ladder, everything becomes more competitive. To begin, you will have to compete with your friends. With each step you will enter a more select group of individuals, who do things right most of the time. The individuals in the higher groups are more confident, more enthusiastic, and more fun to work with.

One way for you to gain a competitive advantage is to use your company's resources to advance your own career. Notice how politicians always convince others to promote their cause. It is the same in a corporate environment. The more people who work for your cause, and the more effective the use of a corporation's resources, the better.

Your use of a corporate organization to work on your project enables you to get more work done in a given amount of time. In a sense, it enables you to direct the work of others, even though they don't report to you. In some cases, you will have individuals on a higher level working for you and even selling the concept of your advancement.

Greg Mandville (see Chapter 1) is a master at getting others to work on his projects. They are caught up in his enthusiasm and help promote his projects. They also helped Greg become a vice president of a major company and learned a valuable lesson at the same time. They learned how Greg's consideration for others and his enthusiasm carried Greg and his associates up toward the top of their company.

6.2   USING THE PATENT DEPARTMENT

A corporate patent department can help you advance your career. For example, you can obtain a wealth of technical information by requesting a state of the art search in your area of technology. Patent searches are also useful for avoiding future patent problems and for helping you to get the broadest patent protection possible.

In general, a patent attorney can advise you on evaluating the patents disclosed by a search. He will also assist you in defining your invention and selling your project. In addition, his opinion can be a valuable asset in promoting your career.

Part of your job is to monitor a competitor's activities and to learn which companies are pursuing an interest in your field.

The assistance of a patent attorney is also useful for obtaining information about a competitor. Ask him to initiate a watch service in the European patent office. Copies of recently published applications will disclose those concepts a competitor feels are commercially significant. You may also learn about new competitors in your field.

Your knowledge of competitive activities could avoid the financial disaster of introducing a new product that is obsolete.

Ted Welton is alert to competitive information and is a master at using such information to promote his own projects. When the patent department called his attention to a competitor's activities, he set up a meeting with the company's president. As a result, he obtained funding for a major program designed to overtake the competition.

6.3   USING THE LIBRARY

Ask your company's librarian to send you copies of the professional journals in your field. If your company doesn't have a librarian, use your public library or one at a nearby university. If your company's library maintains a clipping service, ask to have articles on leading competitors or potential competitors sent to your attention. You can also ask the library to conduct a literature search for any article pertaining to your project. Maximum utilization of these resources can optimize your use of your time.

6.4   MORE ABOUT PATENT SEARCHES

Use patent searches to obtain information for your project. A novelty search is inexpensive and can be used to determine whether your idea is patentable. This type of search provides a brief glimpse of the patent art and indicates how broad a patent you can expect to obtain on your idea.

At times, you will want a more thorough search, which will disclose any patents that might be infringed by manufacturing a product based on your invention.

Neither of these searches will disclose pending patent applications, since applications are maintained in confidence by the U.S. Patent Office. Therefore, you cannot be certain that someone else does not have a patent application that would anticipate your idea or present an infringement problem. You can use this fact to justify urgency in working on your project and filing a patent application as soon as possible.

Engineers frequently overlook the effect of design changes on patent considerations. A search conducted on your original idea might not disclose a patent that would be infringed by a modified design. Therefore, as an astute engineer, you should ask to have a search brought up to date as your product approaches the commercial stage.

Each week the U.S. Patent Office publishes the *Official Gazette*, which includes a drawing and initial claim of each patent issued. Your attorney will, at your request, send you copies of each patent that issues in your area of interest. He can also include copies of patents issuing to competitors.

Another type of search, an assignment search, can identify the patents of a competitor or a leading scientist. You can benefit from reviewing their progress and advising others in your company about competitor's potential new products. Analyze the competitor's work and assess its value.

6.5   BUILDING YOUR SALES FORCE

A team of salesmen can sell far more effectively than an individual. Your job is to develop a team of salesmen who will work at advancing your career. Developing this team is not difficult, but does require a conscious effort.

Successful lawyers, politicians, and professional salesmen get their names in front of as many people as possible. A car salesman wants his name to come to mind whenever someone considers buying a new car. You want your name to come to mind whenever anyone is considering offering a promotion.

You should send interesting articles, patents, or information about competitors to your superior. Articles from the *Wall Street Journal* and *Business Week* should be distributed to everyone who may have an interest in the article.

Read your local paper and send copies of any articles on associates to them. Everyone likes to be noticed and will appreciate the extra copy. Be alert for articles on their children and congratulate your associate on his children's success. Being considerate of others goes a long way in establishing your sales team. A few words of praise or a "thank you" will earn added cooperation and an ally in promoting your career.

Bill Wattmore, a retired executive vice president at Carborundum, had a reputation with the company's engineers for being anti R&D. At the end of one presentation, he asked when he could expect to get his investment back and didn't get an answer. It's a matter of learning what the customer wants. Had they given Bill what he wanted, they could have developed an ally.

Serving on committees is another way of obtaining professional recognition and establishing yourself as an expert. Copies of committee reports can then be forwarded to your associates.

Work with your superior, not against him. Your supervisor can do more toward advancing your career than almost anyone else in the company. Outstanding performance appraisals, comments to his boss, and high visibility assignments are all within his prerogative.

Your supervisor is one step above you on the corporate ladder and has a different perspective. Listen to his advice. Be sensitive to what he says and do everything in your power to make him a success. Sell your superior in the same manner that you would like him to sell you.

6.6 USING THE MARKETING ORGANIZATION

Central R&D is frequently criticized for its lack of market awarness. Your company's marketing group can help you to become more aware of your customer's needs and can provide an important lesson in marketing concepts.

Learn all you can about marketing and apply it to selling your projects and yourself. Lunch with a marketing professional will help you more than an incestuous conversation with your colleagues. Ask questions about market share and price elasticity. Find out which of the company's products do well and what marketing needs to do a better job. Send copies of articles from technical journals to your friends in marketing and exchange information on competitive analysis.

Ask your colleague in marketing how he would sell your proposed product. If, for example, you have developed a longer life grinding wheel and a customer is interested in metal removal per hour, sell your development on that basis. Ask your colleague to recommend several basic books on marketing. The more you learn about marketing, the more you will do to advance your career.

Look for opportunities to work with marketing and analyze the marketing efforts of your competitor. If competitors use misleading technical information in their ads, call it to marketing's attention. You should also call such ads to the legal department's attention. In addition, suggest both ways that marketing might use technical information in their programs and additional applications of your company's products.

Ask to participate in a sales call, particularly if your company's product is encountering problems. Clear this with your supervisor first and recognize that participating in a sales call can provide an exceptional learning experience. Make certain that you listen attentively and follow your salesman's lead.

Dr. Dean Batha, director of development at Fiber Materials International, believes that "an engineer or PhD ought to spend at least one year in a marketing organization." He recommends that "they should leave the (R&D) field after 3–5 years which would then be an appropriate time to get marketing experience."

6.7   USING THE PERSONNEL DEPARTMENT

The personnel department can also help you to do a better job. Its members were instrumental in hiring you; your success will be a favorable reflection on their efforts.

Ask the recruiter who hired you about courses offered by the company. Ask him to recommend good books on human relations and a short course on leadership development.

Let personnel know that you are interested in doing a better job. Find out what they look for in appraising an individual for a higher position. They can frequently suggest ways to improve your chances for promotion.

Ask personnel for the company philosophy on continuing education. Will your company pay for an M.B.A. or a graduate degree in engineering? Are there any books on company sponsored courses that will help you to prepare for a promotion? Follow up with personnel after reading the suggested books or taking a course. Send them copies of articles on managing professional engineers.

You can also offer your services in helping to sell the company to new recruits. As a recently hired engineer, you can answer a recruit's questions and allay fears that might not be raised with a supervisor. You can also learn more about the personnel function by working with them to attract high quality employees to your company.

Be sincere and personnel will help you to get ahead.

6.8 USING THE FINANCE DEPARTMENT

"There is nothing wrong with greed as long as it is tempered by avarice," according to L. M. Foley. In essence, management funds projects that offer the greatest return with the least amount of risk.

Your job is to have corporate resources allocated to your project. Getting money for your project takes ingenuity and salesmanship. Remember that the most effective salesman finds out how to satisfy a customer's needs.

Analyze your project in financial terms and apply discounted cash flow methods in selling your project. Talk to the financial analysts and obtain their help in learning the fundamentals of their profession. Read one or two books or attend a course on financial management in R&D.

Talk to someone in the finance department about your project. Find out what types of questions are raised and get the answers. Don't be embarrassed if an executive asks when he will have his investment returned.

6.9 DEVELOPING A TEAM

Greg Mandville always has help on his projects. The fact is that everyone enjoys working with him. Greg learned early to utilize other organizations.

He involved the patent, legal, and tax departments on a proposed joint venture or acquisition. He also involved engineering and finance, orchestrated the group, and accomplished things in an incredibly short time.

Greg also involves management early in the planning process and keeps them fully advised throughout, so there are no surprises.

You may ask where you will get the time to involve so many people and organizations in advancing your ideas and your career. You may wonder if you will have sufficient time to be an engineer and to do the work for which you are paid. If you use your time efficiently and avoid wasting it, there will be more than enough. Besides, each of the people you talk with can help you do a better job and sell your promotion.

6.10   EFFECTIVE USE OF YOUR TIME

One impediment to success overshadows all the others, procrastination. This habit can and will destory your career, unless you overcome temptation to procrastinate. George Michael (see Chapter 5) wouldn't start a job because it was too close to coffee break. He put jobs off until after lunch or until tomorrow.

Start now by breaking each of your jobs into manageable portions. Start today and get something completed. Coffee and smoking are both bad for your health and worse for your career. Don't waste your most valuable commodity or destroy a promising career by taking time for frequent breaks.

Everyone in your company has the same amount of time in each day, week, month, or year. Nevertheless, it is not necessary to become a grind or to work 16 hours each day. However, if you want to succeed, it is necessary to work intelligently and to utilize short periods of time productively.

If you waste 15 minutes each morning and another 15 minutes in the afternoon, you will have wasted 3 weeks of full time effort in a single year. By using the 3 weeks to complete minor jobs, you will develop a reputation as a doer and as someone who gets his jobs done.

Have you noticed the top executives in your company. If they are like most senior executives, they are constantly busy. To a large degree, it is a matter of self-motivation and self-discipline. These same executives are looking for young professionals with the same disposition for promotion.

Henry Hemenway, president of Industrial Experiences, is always active. Henry retired as a group vice president of Carborundum, but continues an active life. Like many other top executives, he doesn't waste time.

In the competition for promotion, you cannot take time to look at the other guy. Don't be like a race driver who is so concerned about the car behind him that he spins out and loses the race. If a competitor gets promoted, so what? It means that he did something right. Concentrate on working efficiently and effectively, so that you catch up and pass him as you climb the corporate ladder.

6.11   ESTABLISHING GOALS

In addition to your career goals, it is helpful to establish daily goals for more effective use of your time. Write down a series of goals each day and see how many you can accomplish. Complete one before starting the second. Plan your day and concentrate on high priority jobs. If you have 10 minutes before lunch, clean up a low priority job like dictating a short memo.

You can handle low priority jobs as expediently as possible. A quick phone call may be all that it takes. Use these jobs as fillers for those times when you have a few minutes to spare.

Using your time wisely is a prerequisite for the transition into management and setting goals is the first step in gaining control of your time.

7

Preparing for
the Transition
into Management

*The elevator to success is out of order, you have to take the stairs one step at a time.**

*Girard, Joe, *How to Sell Yourself*, page 200. Copyright © 1979 by Joe Girard. Reprinted by permission of Simon & Schuster, Inc.

If you apply management techniques in your present position, you will increase the likelihood of promotion and will prepare yourself for a smooth transition into a higher level of management.

Henry H. Hemenway, president of Industrial Experience of Worchester, Massachusetts, says that "to succeed in engineering management you must understand two principles."

The first principle is to understand the value system of your company. "Too many engineers don't question total costs," according to Henry. He maintains that "high operating costs or frequent service calls can destroy your business." "In high technology companies you must be at the leading edge of technology but costs are even more important," he adds.

"Understanding the business which your company is in is the second principle," according to Henry. He suggests that "a star can't go any place in a stagnant company." Henry also says, "It is better for a company to be No. 1 in a specific area rather than second or third in a number of markets."

Sam Marcus, director of human resource development at AMF, says, "Engineers should take inventory of their talents." Sam believes that "engineers have a rare skill, an ability to apply a scientific approach to problem solving." He advises you "to apply that ability outside of your technical discipline."

7.1   PLANNING

How well you plan in your present job is an indication of how well you will plan for a large group.

Your first step is to get control of your time. Establish a plan for each day and try to follow it. Plan for tomorrow before leaving the office and generate enthusiasm for the next day. List the projects that need work, itemize the tasks, and rank them in order of importance.

When you plan the work for a technician, make certain that you clearly explain the activities. If he fails to understand the importance of each step, you may end up doing his job.

Walter H. is an extremely bright, hardworking scientist. At a major electronics company, he had difficulty getting cooperation from the technicians and claimed that they failed to follow his directions. He resorted to doing everything himself and was a "hands-on" physicist. Had he put more effort into planning the activities of his technicians, he could have accomplished much more. He failed to demonstrate management ability and returned to the academic life.

Don't let crisis rule your life. Good planning can help you to take crisis in stride. You can determine whether the crisis is more important than the project you are working on. If not, finish the present project before diverting your attention to the new challenge.

There is another problem associated with crises; one or more crises can cause you to loose perspective. You may become so busy that you forget to take time for planning or neglect a well-thought-out program.

Tom Fowler, a foreman at Western Electric, did an exceptional job. Tom was constantly moving and gave the appearance of being crisis oriented. In production, he responded to any crisis that might interfere with meeting his program, but he also took time for planning. Tom was promoted, applied his principles of planning, and overcame a major production problem that threatened to shut down an entire plant. Tom Fowler succeeded because he took time for planning. He also taught me a number of lessons, for which I shall always be appreciative.

7.2 REVIEW YOUR CAREER GOALS

Planning means more than gaining control of your present job and preparing for a promotion. Career goals should be reviewed at least once every 6 months. Ask yourself what you have done really well in the area of career advancement. Determine where you could have done better and resolve to improve in those areas during the next 6 month period.

Henry Hemenway advises young engineers to "determine what you want to be ten years from now."

Robert Townsend, former president of Avis Car Rental and author of *Up the Organization*, suggested that no one should stay in the same job for over 5 years. There are exceptions to the rule, but consider it in measuring your own progress.

"The worst thing a technical expert can do is to stay in a staff position," according to Bruno Miccioli. This means that you must plan for your own promotion and, even more important, that it is up to you to make it happen.

When you review your career progress, remember what Henry Hemenway said about understanding your company's business. Recognize that companies and jobs change, particularly with changes in top management. Nevertheless, if a company is stagnant or if you don't feel comfortable with management, it may be time to move.

If you are dissatisfied with your job, continue to plan. Utilize your time effectively and do the best job possible. Good work habits are transferable to another job. You will also be remembered as a hard worker in case you need a reference in the future.

Consider that in many companies even the technical drones get routine raises and that the differences between routine raises and above average raises is usually small. Some companies even adopt a system that routinely rewards mediocrity and many superiors go along with the system.

The Hay system is characteristic of management's approach to evaluating job performance and to establishing salary ranges. Each position is assigned an arbitrary number of points based on educational requirements, interaction with higher management, impact on earnings, and the like.

The Hay system and others like it allow management to equate two and one half controllers to one brain surgeon. Therefore, if you work for a company that uses this type of system, it may be like working for a manager that has had a lobotomy. For example, one salary survey explained Hay's popularity. This survey showed that, on the average, companies utilizing this type of system paid 20% less than those that did not. Is it any wonder that companies paying mediocre salaries employ mediocre people? Whatever else you do, don't make any long range commitment to a company with a philosophy of paying mediocre salaries.

Fortunately, progressive companies and managers recognize the importance of rewarding creative individuals. They find ways to reward extraordinary performance with meaningful raises and promotions. By working hard and directing your efforts, you will obtain meaningful raises and promotions.

7.3   ORGANIZATION

A cluttered desk is not a sign of genius. For example, you can keep track of
your projects and complete them on schedule if you plan carefully, stay
organized, and see that each job gets done. Follow your list of things to do,
use a PERT chart, and make certain that your job doesn't delay a larger
program.

Take advantage of small increments of time to control the progress of
your project. Use a few minutes to follow up on work in the model shop or
to keep your superior advised of your progress.

Integrity is your most valuable asset. You can protect your reputation
for integrity by being organized and by giving a full day's work for a day's
pay. To do less is to cheat your employer. After all, you are paid for what
you do and judged by your accomplishments. Don't join the group for a
long coffee break or waste small increments of time for a cigarette.

Don't make promises that you can't keep. If a schedule is unrealistic,
say so. Tell management that you are concerned about meeting the
proposed schedule and explain why. Don't be guilty of telling management what they want to hear and then damaging your reputation when you
can't produce.

Cost control can be almost as important as time control. Don't ask for
$10,000 if you need $50,000. Most managers recognize the difficulty in
preparing engineering estimates and will appreciate prompt notification of
any deviations from budget. Be candid and protect your reputation for
integrity.

You can control your destiny by controlling your time and your work.

7.4   RECORDS MANAGEMENT

Every engineer should keep good records. You will help your employer
and yourself, and earn the gratitude of your patent lawyer, by enhancing
your company's prospects for prevailing in a patent dispute.

Keeping records is not difficult and does not take a lot of time. You
should use an engineering notebook; a simple notebook will do. At the end
of each day, take a few minutes to note your accomplishments, then sign
and date each entry.

Your record keeping should begin before you make an invention. Use your notebook to keep a log of your daily accomplishments and ideas and sign your name and the date below the description. If you think that you may have made an invention, have your entry witnessed. Ask the witnesses to write "witnessed and understood" before signing and dating the entry.

Write legibly and use a pen to produce a permanent record in your notebook. Describe each of the essential elements of your invention. When you make an entry in your notebook, write in a manner that can be understood by a technically trained individual without any need for further explanation.

Do not erase or leave blank spaces in your engineering notebook. Don't go back and change an earlier entry. Such practices can destroy the value of your records. If subsequent tests disprove an earlier record, record that fact on the date of the subsequent test.

Record your progress in developing an invention and, equally important, explain any delays. For example, if your project is sent to the model shop and encounters a month's delay, record this fact. Otherwise, in some future legal proceedings, it may be argued that you were not diligent in pursuing the development work or that you abandoned your invention.

A "reduction to practice" refers to a first working model or trial use of your invention. You should describe a reduction to practice, as well as each of the elements incorporated in the working model, in your engineering notebook. Don't forget to sign your notebook and have the entry witnessed and dated. You should also treat subsequent tests with various modifications in a similar manner.

Your records will determine your rights to an invention and will be the subject of careful scrutiny if your patent is involved in litigation. Therefore, avoid statements that could prove embarrassing if read by others and those that might be difficult to defend under cross-examination. Avoid opinions indicating that your invention is inoperative or very limited.

In summary, record the facts legibly and permanently. Then ask your patent attorney to review your records and to make suggestions on how to improve your practice.

If you visit a customer or vendor and make suggestions on modifying their practices, record that fact in your notebook. Some years ago, a young engineer taught a customer how to improve their process by incorporating a material manufactured by the young engineer's employer. The customer obtained a patent and prevented the engineer's employer from using the invention. They could do this because the young engineer had failed to record his idea in his engineering notebook.

Ask your secretary to keep your records (other than your engineering notebook) in an appropriate file clipped in place in chronological order. Her desk calendar can then be used to indicate when it will be needed. Your secretary can also keep a 3 × 5 card file to assist your memory; however, it is probably better to keep your own. Vendors' and customers' cards with personal information, such as birthdays and interests, will refresh your memory before a subsequent meeting. Include each person's coffee preferences and impress them on subsequent visits. In addition, you will develop a reputation as a person with an excellent memory.

If you can't take notes during a meeting, sit down as soon as it is over and record your thoughts. Reviewing those thoughts before a subsequent conference will help you to control the situation. It will also free your memory for more important information and add confidence in your own ability.

Ted Welton carries a small stack of yellow cards about 2½ × 4 inches with his name at the top. He jots down a short request or reminder for an associate or for himself. These memory joggers help Ted to do a better job and keep his associates from forgetting to follow up on a request.

The nice thing about Ted's memory joggers is that you can keep them until the job is done. When it's done, dispose of the note and with it will go the worry that you have forgotten something important.

7.5 THE SELF-DISCIPLINE OF TIME SHEETS

If you keep an accurate account of time, you will become more productive and will do a better job.

Joseph Cesarik, a corporate lawyer at Sohio Industrial Products Company, is an aggressive lawyer with a fetish for keeping track of his time. Like many lawyers from private practice, Joe accepted the principle that if you don't account for your time, you won't get paid.

For several years, I watched Joe and continued with my own hit or miss effort at time-keeping. One day, I decided to follow Joe's example and a strange thing happened—I noticed a dramatic increase in productivity. After several months, I learned that productivity continued to increase as I accounted for smaller increments of time.

Today, I keep an open diary and a small clock on my desk. If a phone call takes 5 minutes, it is recorded with a brief description. If a patent application is discussed for 45 minutes, the time is accounted for. If you make a personal phone call, record it. In practice, I have found that 10 to 15 minute intervals are most effective for me. Several short phone calls can be lumped together in a 10 to 15 minute interval.

After reviewing your day's work, can you say, "I really had a day to be proud of." At times, you may encounter a less than satisfying effort. Write it off by resolving to make up for it tomorrow.

7.6 GETTING STARTED

A retired patent lawyer said, "If you don't get up in the morning, everything else is null and void." Getting up means more than dragging yourself out of bed. Psyche yourself up to meet new challenges. Today could be the day you earn a promotion.

If you have trouble getting up, start off with a shower. Clean out all the cobwebs from your mind before leaving for the office. Tell yourself that you are a winner and take charge of your day. Remember, you can only go through today once. Do it in a way that will make you proud.

An unknown philosopher said, "Today is the first day of the rest of your life." How today goes will set the course for your future.

You can set the stage for a successful day by ending each day on a positive note. Complete the entries in your engineering notebook and reflect on your accomplishments. Clear off the top of your desk. Don't dump everything in a drawer. Get things organized, so that tomorrow you can start right in on something exciting.

If you still have a few minutes left, clean up one of those little jobs—dictate a short memo or fit in a couple of phone calls. If you are in the

East, place the call to the West coast instead of putting it off until tomorrow. You'll feel better knowing that another task has been completed.

You can also take the last minute or two before leaving the office to thank God for his help and to ask Him to help you to do a better job tomorrow. Try it, it works.

7.7 GETTING INVOLVED

Professional organizations can help you in making the transition into management. Working on committees and seeking office will help you to develop management skills. Getting to know people and learning how they have solved similar problems will make you more effective.

Sam Marcus, director of human resource development at AMF, says, "Develop sensitivity to people by putting yourself in the shoes of the other person." Sam also suggests that "you should develop a broader perspective—break free of your specific discipline and develop a balance between analytical and creative appraoches to problems." He adds, "Become well-rounded through literature and the arts."

You can win recognition by presenting papers or joining in committee work and by sending copies of these reports to others to keep your name in the limelight. Joe Girard, in *How to Sell Anything to Anybody*, emphasizes the importance of keeping your name in front of as many prospects as possible. Remember the importance of selling yourself.

Establish your credentials as an expert by joining in professional activities and thus make yourself more attractive to other companies. Executive recruiters use professional activities as a source of candidates who are ready for a more challenging opportunity.

8

Establishing
Your Rights

You should take the proper steps to protect your ideas and to establish your rights.

Progressive companies are vitally interested in your creative ability. Many of them have an in-house patent department to protect your ideas and to help the company obtain a reasonable return on its investment in your inventions.

Some smaller companies establish a working relationship with an outside patent law firm for the same reason. Under this arrangement, a patent lawyer makes periodic visits to their facilities, but he is probably less effective at ferreting out ideas than an in-house attorney. In these cases, you should seek out the attorney and promote your own ideas.

If your company's executives are not patent conscious, you have an opportunity to sell them on the importance of patents. You can show them how an investment in a patent can be used to increase their market share and the return on their investment. In establishing your rights, don't sit back and wait for an idea to be discovered. Find out the correct procedure for obtaining a patent evaluation, initiate a patent search, and promote your ideas.

As you know, a good patent lawyer can obtain a patent on almost anything. Nevertheless, you should ask yourself if a patent is cost justifiable. Cost justification is determined by evaluating the breadth of the proposed patent claims in light of the market potential. Will your patent give your company a competitive advantage? If so, you can use this fact in promoting your ideas.

A number of your ideas are probably patentable and can be sold on their merits. Nevertheless, emphasis on the market potential will help in convincing management of the importance of patent protection.

8.1 LOOKING OUT FOR NUMBER ONE

Your idea has very little value until you sell it to your management. Submitting your idea to the company's patent department is only an early step in selling an idea. However, the patent department can help to sell your idea and aid in obtaining recognition for you.

Start by selling your patent laywer on the commercial potential for your idea and enlisting his aid in getting the broadest possible protection. It may require perseverance, since some companies have reduced the size of their patent staffs and have restricted patent filing to cases scheduled for commercialization. However, exceptions are made for pioneer inventions and in those cases where the inventors are persistent.

Professor James Bright, speaking to the Licensing Executive Society, reported that important inventions take 25 years or longer to reach commercial success. Frank C., treasurer at a major industrial company, commented that with today's high interest rates such long term projects cannot be justified.

Recognize the interests of your company and emphasize projects that meet their criteria. You can succeed in convincing management to invest in a long range project. To do so, you may have to overcome an investment criteria of a 1 to 2 year pay-back. This means that you must convince someone in top management that your idea has tremendous market potential and a high probability of success.

If your interests lie in longer range projects, you have other alternatives. Be honest about the risk, but point out the long range potential and suggest a joint venture as a possible means of reducing the risk and the size of the investment. This approach almost guarantees the attention of higher management and is discussed in more detail in Chapter 13.

A second alternative for selling your long term projects is to seek government funding, which is treated in Chapter 10. This approach usually requires that you sacrifice a portion of the patent rights. Nevertheless, you should point out all of the alternatives to management and let them make the decision. It is far better to obtain government funding than to abandon your idea.

Remember that each time you attempt to sell an idea, you are selling yourself. Give management the facts, the alternatives, and your recommendations. If they elect to discontinue work on your idea, they should recognize a well-thought-out program. You should have impressed them with your analysis and presentation. If all else fails, seek a release to pursue your invention on your own. This is a drastic step and may have an adverse effect on your career. Nevertheless, like management, you should consider all of the alternatives.

8.2   BECOMING A TEAM MEMBER

Recognize the contributions of others as you would like them to recognize yours. Bob Meltzer did this and built a successful team of researchers based on mutual respect. The members of his group were open in discussing their ideas. Frequently, they embellished the ideas of one another and with Bob's market orientation were successful in selling management on their projects. Bob went on to become a vice president.

By contrast, there were two individuals at Western Electric who avoided the team concept. If they came up with an idea during a conversation, they would race back to their desk and prepare an invention disclosure. They were afraid that someone else would get the same idea. Both were prolific inventors, but neither was promoted.

In some cases, you may make a basic invention and, because of the synergistic effect of two minds working together, may become a coinventor of an improvement to your basic idea. The synergism provided by two minds frequently makes the difference between an interesting invention and a commercial product.

8.3   INTEGRITY IN THE PATENT OFFICE

You should be absolutely candid in dealing with your patent attorney. Advise your attorney of any offer to sell a device that is even similar to that contemplated by your invention. Show him all your test results, including those that indicate it won't work.

Some years ago, an inventor at a major institute made an important invention. His attorneys were preparing to argue his case before the Court of Customs and Patent Appeals, because they believed he was entitled to broad claims. Several days before the appeal, one of the attorneys reviewed the inventor's engineering notebook and uncovered a serious problem. The examples that were the basis for the broad claims were listed in the notebooks as "NG." The broad claims were cancelled.

You can protect your company from an adverse judgment by being candid with your patent attorney. For example, a patent lawyer for the Norton Company of Worcester, Massachusetts failed to call a pertinent prior art reference to the attention of the U.S. Patent Office. A U.S. district court decided that this failure was fraudulent and awarded $450,000 to the other side.

Typical costs for patent litigation range from $300,000 to $500,000 or more for each party. A good patent on your idea will justify this investment. However, a patent tainted with fraud (failure to disclose your knowledge to the Patent Office) can expose your company to added costs and the loss of what could have been a valuable patent. Besides, your reputation is at stake.

8.4   SUBMITTING AN INVENTION DISCLOSURE

If you think that your idea might be patentable, call it to the attention of your patent department. Time may be of the essence.

There is no magic formula for submitting your invention for a patent evaluation. However, you can help the corporate patent department to do a better job by writing a clear description of your idea. Many companies provide an invention submission form such as the one shown in Appendix B. These highlight the type of information that is helpful in evaluating your disclosure. In submitting an idea, define your invention and describe how it works. Be sure to include all the essential elements for a working device. For example, a submission of "an inspection device for scanning the interior of a pipe" would not be sufficient for a patent evaluation, unless you describe the essential elements and how the device works.

A brief description of the state of the art and how your invention differs from commercially available equipment will help your patent lawyer. Remember, he may not be familiar with the technology and will appreciate background information to help him understand your invention.

In writing a disclosure for your invention submission, describe your idea in simple English and avoid oversophistication. An average patent lawyer does not have a Ph.D. and will usually appreciate the use of lay terms. Besides, a clear disclosure will help your attorney to obtain a good patent search and to distinguish your invention from the prior art.

8.5   THE FUNDAMENTALS OF INDUSTRIAL PROPERTY

Knowledge plus perseverance equals power. Your knowledge of industrial property plus perseverance will help you to obtain patents on your ideas. With knowledge, you will become more effective in dealing with lawyers, in promoting your ideas, and in getting bigger raises and promotions.

Don't be discouraged if nothing happens after submitting one or two invention disclosures. Don't assume that no one is interested in your ideas and give up on inventing as a means of profit or a key to advancement. The problem is that many corporate patent departments are deliberately understaffed to restrict patent activities to projects that have the highest commercial potential.

To avoid this problem, ask for assistance. When you submit an invention disclosure, ask for a novelty search. If your invention is a basis for a major project, ask for a state of the art search. Point out the commercial importance of each idea and express a degree of urgency.

If your idea is important to you, follow up on your submission. Find out its status in the patent department and how soon you should expect results.

As you work on new ideas, learn how to recognize patentable features and to distinguish your ideas from those disclosed in the prior art. Learn to work with a patent attorney. Go to lunch with your patent lawyer and discuss the basics of novelty, validity, infringement, interference proceedings, litigation, and licensing.

In the words of Napoleon Hill, "The man with a definite purpose will take careful inventory of every person with whom he comes in contact in his daily work, and will look upon every such person as a possible source of useful knowledge." He goes on to say that "his place of daily labor is literally a schoolroom in which he (you) may acquire the greatest of all educations."*

In reviewing the fundamentals of patent law, you have already learned that your idea is probably patentable. Nevertheless, a patent examiner will probably reject your patent application, possibly because the idea is not new. In patent terminology, this is referred to as a 102 rejection. The examiner may also reject your application because your idea was obvious to a person of ordinary skill in the art (referred to as a 103 rejection). In general, the examiner will combine two or more references to show that the invention would be obvious to a person of ordinary skill. At other

---

*Hill, Napoleon, *The Master Key to Riches*, page 96.

times, your patent application may be rejected under section 112, which means that the claims are indefinite or too vague.

Your attorney should discuss any rejection with you, as well as any arguments to be used in overcoming the rejection.

If your patent application is rejected, don't be discouraged, since many rejections of patent application are overcome fairly easily. Nevertheless, you should consider the validity of your patent. As previously mentioned, a patent examiner has a limited amount of time to consider an application and is more lenient than a court in considering the level of invention required for patentability. You and your attorney should work together to obtain the broadest patent possible. However, don't fall into a trap by arguing for broader protection than you deserve.

At times, your patent application may be involved in a patent interference, a patent office procedure to determine which of two or more applicants is entitled to a patent. These procedures are complex and depend on written records to determine the first inventor.

8.6   USING A PATENT

There is nothing wrong with trying to put a competitor out of business by obtaining a strong patent position. The Xerox Corporation and Polaroid both succeeded in keeping competitors out of their fields for many years by establishing legal monopolies.

In considering your patent application, look at the claims to see if your claims will keep a competitor out of the business. If a competitor can avoid your claims by making minor modifications, you should question the value of the patent. Nevertheless, because of the high cost of litigation, some companies will respect your patents even though the patent appears narrow or invalid. Such companies will be reluctant to copy the results of your research and development.

On the other hand, as an aggressive engineer, you may want to copy a competitor's product without involving your company in litigation. A review of the Patent Office records may help. The Patent Office's records include all of the arguments an applicant used to get his patent. These arguments can be used to limit the scope of the claims and are referred to as "file wrapper estoppel." Your attorney can review these records and advise you about the risks of litigation.

A patent is presumed to be valid, which means that an infringer has the burden of proving invalidity. Therefore, if you want to copy a competitor's patented product, ask for an extensive patent and literature search. Have these searches extended into the foreign art and even review doctoral theses to show that their patent is invalid. Because of these extensive searches, the courts find that about 75% of the litigated patents are either invalid or not infringed.

Nevertheless, don't underestimate the value of a patent. The alternative to seeking patents is to give the results of your research and development to your competitors. Besides, you can increase the likelihood of getting a strong patent by conducting an extensive state of the art search before filing a patent application. If you call the most pertinent references to the examiner's attention and include claims of varying scope, you can obtain a strong patent and make it more difficult for a competitor to invalidate your patent.

One approach is to include narrow claims in a patent application, since these are more difficult to invalidate. Narrow claims may even provide adequate protection on a commercial product.

In some cases, you may find a more pertinent reference after the issuance of your patent. Fortunately, there is a remedy. Under certain cases, your patent claims can be changed by filing for a reissue of the patent. However, an infringer may avail himself of a "doctrine of intervening rights," which permits him to continue what he was doing before the reissuance of the patent.

You should discuss with your attorney detailed questions regarding interferences, the doctrine of equivalence, and intervening rights. Your basic understanding of these concepts will enable you to utilize the legal staff and to position yourself for a promotion.

8.7   SELLING A PATENTABLE IDEA

A team effort is usually more successful in developing a new product than an individual effort. However, some inventors refuse to accept the team concept and condemn their ideas to failure. Once you recognize the value of your idea and the role of patentability in selling your idea, get help. You will have shown that your idea is patentable, enlisted the aid of a patent lawyer, and demonstrated that patentability increases the potential value of your idea.

By this time, you may have a working model or, in the language of a patent lawyer, a reduction to practice. You should have evaluated various alternatives and tested the more promising variations. In effect, you should have established the technical feasability of your ideas, provided for their legal protection, and prepared a more effective sales presentation.

By reducing a number of alternatives to practice, you can improve your company's patent position and illustrate the various applications for your idea. It is also likely that you will make further inventions and will be able to patent the improvements. Actually, you can develop a network of patents that will prevent a competitor from entering the field.

For example, some years ago, a client company was considering the acquisition of a foreign company to produce glass retroreflectors for highway signs. The 3M Company had over 30 patents that seemed to cover most of the practical alternatives. Because of this network of patents, my client decided against entering the field.

In selling your idea, utilize the help that is available from a corporate organization. Larry N. is a brilliant scientist and an entrepreneur. Larry utilized the patent staff, but refused to accept the help of a mechanical designer in developing his invention. In fact, he became dissatisfied with a corporate career and formed his own business.

Take advantage of your company's experts. The marketing organization can provide useful information and can help you to sell your ideas. Ask how they would sell management on investing in your development. Ask someone in marketing to allow you to observe or help when he is preparing for a formal presentation. Note the use of visual aids, slides, and graphs. Marketing professionals know how to sell. It's a skill most engineers lack and one that is relatively easy to learn. It can make the difference in a major promotion.

Also, don't overlook the company's public relations organization. If you have written something about your idea, ask for their assistance. If you are called on to make a formal presentation, write it and ask for their editorial comments. Your professionally polished presentation will make a favorable impression on management.

9

Establishing
the Value of an Idea

9.1   HOW TO WORK EFFECTIVELY WITH A PATENT LAWYER

A patent lawyer will do most of the work in preparing and prosecuting your patent application. Nevertheless, it takes a team effort to obtain the best possible patent.

Your job is to explain your invention in a manner that can be clearly understood by your attorney. Describe your invention, point out its essential features, and suggest alternatives for each feature. Describe any optional features and explain how the invention works. Suggest other forms and applications.

A description of any competitive products and the advantages provided by your invention will also help your attorney to prepare your patent application.

Howard J. is an excellent patent lawyer with extensive experience in mechanical engineering. He obtained a patent on a chemical process that included chemical treatment and washing the treated product to remove excess chemicals. The inventor failed to suggest that excess chemicals could be removed by a centrifuge. Better claims were obtained after the inventor understood his role in obtaining broad protection.

Be candid with your lawyer. Tell him about any offer to sell a product based on your invention or any similar product manufactured in the past.

Leslie O. Vargady, an entrepreneur, is one of my favorite inventors. Some years ago, after reviewing a patent application I had written, Leslie said that the claims were too broad, since they would cover a device manufactured in 1937 in Germany. If Leslie had supressed the information, we would have obtained broder claims, but they would have been invalid.

Some corporate patent lawyers establish their own priorities. However, they are subject to supervision, personnel appraisals, and management by objectives programs. You may, by emphasizing your idea's commercial significance, convince your patent lawyer to give your application a higher priority. You can also advance your cause by being cooperative.

Another approach for advancing your idea is to enlist the aid of your boss. Have him call your patent attorney to inquire about the status of your invention and to comment on the commercial importance of your idea. At times, it may even be appropriate to have your boss call the head of the patent department.

Mike Zall, a senior patent attorney at AMF, works with the presidents of business units in establishing priorities for filing patent applications. A supervisor's enthusiasm for an invention will influence the head of the business unit. Mike's enthusiasm is also part of the equation.

Robert J. Meltzer is enthusiastic about his ideas. Bob is easy to work with, provides clearly written disclosures, and makes himself available for consultation. He sells his ideas to top management and to his patent lawyer.

When Bob reviews his patent application, he asks himself how he could avoid the proposed claims. He applies his technical expertise in seeking claims that are as broad as possible without being anticipated by the prior art. You can benefit by following Bob's example.

You should review the entire application. For example, review the objects of the invention to make certain that they are correct. If your attorney alleges that an object of the invention is to produce a less expensive product, when the product is actually more expensive, this could be construed as misleading the patent office.

Don't follow the example of Francis T., a successful scientist, an inventor, and a leading expert in his field. Francis was always difficult to work with and usually unavailable to explain a disclosure. He was reluctant to review a patent application, yet would not sign his name until numerous changes had been made. After reviewing the changes, he would request more changes that had little or no effect on the patentability. No one wanted to work on Francis's cases. Nevertheless, his inventions were important to the company and were written in spite of the lack of cooperation. However, they were not given priority.

If you publish a paper that discloses your invention, or offer a product for sale more than 1 year prior to filing a U.S. patent application, you are barred from obtaining a patent.

In many foreign countries, you are barred from obtaining a patent even if you publicly disclose the invention before filing your original patent application. Therefore, inform your lawyer before you submit any material for publication or before discussing your invention with anyone outside the company. Alert your lawyer before anyone in marketing is permitted to disclose your idea to a potential customer.

Some engineers believe that an experimental sale is permissible without jeopardizing the patent rights. Don't believe it. Experimental sales require documentation and should only be done after obtaining legal advice.

You can improve your working relationship with your patent attorney by writing a clear description of your idea and by requesting either a patentability or state of the art search. If you ask for a patentability search, request at least 6 to 12 references, so that you can learn what others have done.

Study the search report and analyze the cited references. If you have difficulty understanding the references, ask your attorney for assistance. Actually, each reference should include an abstract of the disclosure and a summary of the invention. Reading these two sections and referring to the drawings should clarify what is in the patent specification.

Explain how your invention differs from any earlier patents and point out the advantages of using your idea. Give your lawyer the benefit of your experience and the information he needs to convince a patent examiner that your invention is patentable.

Your goal in assisting a patent lawyer is to be recognized for your contributions.

9.2 WHAT YOU CAN EXPECT FROM YOUR PATENT ATTORNEY

There is more to patent practice than transforming a patent disclosure into a form that is acceptable to the Patent Office. A patent attorney wants to obtain a patent that will be upheld by a court. He knows that courts apply more stringent standards in determining patentability than the Patent Office.

Your patent lawyer recognizes that he is preparing a complex legal document, which will be carefully evaluated by potential competitors, and that a minor error can result in the loss of valuable rights. He tries to write a clear, concise, and complete description of your invention. He will also try to draft claims that cover every conceivable alternative to your idea and that are not anticipated by the prior art.

After drafting an application, your patent attorney will discuss the application with you and correct any errors. He will also ask about contributions by others to determine that the true inventor(s) are named in the application.

If your invention is relatively basic (i.e., a pioneer invention), you may be asked for additional examples to show that you are entitled to broad coverage; a number of examples are required to support broad claims and are also important for chemical inventions.

Before filing your patent application, you must sign a declaration that you believe you are the first inventor and that the product incorporating the invention has not been publicly disclosed or on sale for more than a year. Whatever else you do, don't make changes after signing the declaration, except by formal amendment.

Once your application is filed in the U.S. Patent and Trademark Office, it will be assigned a serial number; after about a year, a patent examiner will review the application and, in most cases, reject it. The examiner will probably argue that it is obvious to combine two or more prior patents as an anticipation of your invention. This is referred to as a 103 rejection and is based on an assumption that your invention would be obvious to a person of ordinary skill in the art, assuming that he saw the prior patents.

At times, you may face a rejection based on section 102 of the patent law. This means that the claims in your patent application read on a prior patent and are, therefore, anticipated.

Your lawyer will send you a copy of the examiner's rejection (an office action), together with copies of the cited references. Some rejections are due to technicalities and can be readily corrected by the attorney. Nevertheless, an attorney should review any office action with you and obtain your agreement on any changes. He should also review the commercial aspects of the invention with you at that time and make certain that the claims in your application cover any contemplated commercial product.

Make certain that your claims cover any proposed commercial products. If a competitor has introduced a similar product, you should examine it and avoid adding any limitation to your claims that would not be infringed by your competitor's product.

Your attorney may conduct a personal interview with the patent examiner and explain the invention and its distinguishing features. An important case usually warrants an interview, since an interview frequently results in the allowance of broader claims. Your attorney may ask you to attend an interview with the patent examiner. In such cases, he should spend considerable time preparing you and should advise you to limit your answers to technical questions.

Some years ago, I was prosecuting an important application that could be explained most clearly in terms of physical optics. One of the inventors was a former university professor and I thought that he could be helpful in an interview with the examiner. After two rehearsals, we agreed that he should not participate, since it was so difficult for him to think like a lawyer. For this reason, inventors are rarely taken to interview a patent examiner.

When a patent examiner persists in his rejection, an appeal may be taken to the Board of Appeals. In such cases, the attorney writes a legal brief and may include affidavits to convince the Board of Appeals that the examiner is wrong and that you are entitled to broader claims. The brief is primarily the work of the attorney, but he should review the brief with you before it is filed.

When your attorney is unsuccessful in convincing the Board of Appeals to reverse the examiner's rejection, he will discuss a further appeal with you. At this point, the costs will be more substantial and economic justification should be considered before filing the appeal.

Your company's patent attorney can also help you in selling your invention. If he is enthusiastic about your invention and proud of his work, this will be evident when he submits his report to management. For example, he may comment favorably on the scope of expected coverage, which will support your argument for going ahead with the project. Your attorney also has contacts with management and may advance your cause with favorable comments about your inventions.

Lawyers who are not full-time employees of your company work in the same manner as in-house counsel, but are usually more conscious of time. They bill the company for each hour spent on company business. They are frequently more conscious of litigation and may devote more time to considering legal consequences. Therefore, in working with outside counsel, you will usually have less direct contact with the attorney.

Some lawyers in both corporate and private practice are primarily concerned about what they can get from the Patent and Trademark Office. This is a shortsighted approach and is characteristic of lawyers with little or no experience in litigation. Your sensitivity to this type of approach could save your company from a costly error.

9.3   FOREIGN FILING ON YOUR IDEAS

About 6 months after filing a U.S. application, your attorney will inquire about filing applications in foreign countries. The cost varies, but typically ranges from $1000 to $3000 per case per country to obtain a patent. In addition, many countries tax patents, so that the costs can become prohibitive for all but the more important inventions.

At a meeting of corporate patent counsels, a number of lawyers concluded that more money was wasted on foreign patent filing than on any other aspect of their operations. Nevertheless, successful inventions are often exploited in the world market and justify extensive foreign filing.

A public disclosure or public use before you file your U.S. patent application will preclude getting a patent in many foreign countries. However, by filing foreign applications within 1 year of your U.S. filing date, you can avoid problems caused by any intervening publications, that is, a public use or disclosure after the U.S. date but before the foreign filing date.

When asked to recommend countries for foreign filing, consider the economics. Think about the delay in reaching the proposed market. For example, a pioneer invention may justify extensive foreign filing. Nevertheless, the time it takes a pioneer invention to become commercial may make it uneconomical to file in foreign countries.

Foreign filing adds to the legal complexity. In some countries, almost any disclosure before filing your original application will prevent you from obtaining a valid patent. Therefore, if you plan on foreign filing, have your attorney prepare a confidential agreement to be signed by a recipient of any information relating to your invention.

9.4   PACKAGING YOUR IDEAS

The value of your idea is directly proportional to your sales effort—it is valueless until it is sold. However, once you convince management to develop your idea, it will increase in value, but the selling job is still not over. An idea must be sold over and over again before it reaches the commercial market.

Paul Marinaccio, director of AMF's Moorhead Paterson Research Center, says, "You have to keep selling a project even after it becomes a successful product. Don't let management forget that you produced a winner."

Selling your idea proceeds in much the same manner as the technical development. With each experiment, you learn more about the idea. The results can then be used in selling the concept.

Report any additional information to your patent attorney as your technical development continues. Your attorney may elect to refile your application to obtain broader claims or to reflect a broader understanding of the invention. In other cases, it may be appropriate to file additional patent applications to cover improvements in your original idea.

Whenever you overcome a technical problem, use it in your sales effort. Be enthusiastic and show how this success brings your idea closer to the market. Don't be preoccupied with technology. Use a market oriented apporach toward technical accomplishments to advance your own career. A market oriented approach will help you in selling your idea and yourself. Be specific and point out how your idea fills a customer's needs.

Corporate executives are interested in a large payout, yet because of financial constraints, invest in projects that promise an early return on their investment. You can, by showing an early return as a step toward a longer range objective, obtain the interim funding for the more important project.

For example, the early developers of holography put considerable effort into the development of medical devices instead of three dimensional television. The market for the latter was thought to be much larger, but required solutions to more serious technical problems.

Whenever possible, ask that a member of the marketing organization be assigned to a development project. They will help with market research and in selling management on its commercial potential, and continuous rapport with members of the marketing organization will help you to identify customers' needs. It should also assist you in making market oriented presentations.

In working with marketing, you should consider an appropriate name for your new product. Suggest several names and submit them to your patent attorney. He will initiate a search to determine if the proposed names would infringe the trademark of some other company. He will also advise you on how to select and develop a valuable trademark.

When you select a trademark, you are identifying your idea as a commercial product. The selection of a trademark also demonstrates your awareness of commercial considerations as well as your efforts to bring your idea to the market as soon as possible.

9.5 SUMMARY

In summary, a good working relationship with a patent lawyer can enhance your image in the company. His enthusiasm for your ideas and involvement in their protection can be an important link in your advancement. Keep your lawyer advised about your efforts and market considerations. Help him to serve you better as you climb the corporate ladder.

10

Recognizing Your Obligations to an Employer

Your obligations to an employer go beyond the duties called for in your written contract. These duties are recognized under the common law and include loyalty and integrity. The Japanese often refer to the spirit of an agreement; this spirit may control your destiny.

10.1 UNDERSTANDING AN EMPLOYMENT CONTRACT

An employment contract usually includes a description of each party's obligations. Why then do so many engineers and scientists treat these contracts lightly and sign them without understanding the contents?

Your personnel department may be somewhat at fault for failing to inform you about the significance of an employment contract. They don't bother to apprise you, because these contracts are routinely processed with other forms, and because most people sign without question.

The fact is that you are at fault if you fail to read, understand, or modify your employment contract. Don't be alarmed, the majority of employment contracts are equitable. Nevertheless, you should read and understand any contract before signing it.

Do not ask your company's lawyer for an explanation. He represents the company and cannot act for both parties. If in doubt, ask your own lawyer to review your contract before you sign it.

If you truly want to do a better job, obtain larger raises, and get promoted, start by reading your employment contract. Don't be intimidated by a printed form, or by a number of forms, on your first day on a new job.

When you accept a new job, ask the personnel representative to send you the company's employment contract, insurance forms, and the like, so that you can review them at your leisure. If you have not done this, I suggest reading them in the evening and returning them by the end of the week.

At times, a company may require your signature or an employment contract before disclosing confidential information. If this is the case, read the proposed contract carefully and delete any questionable terms, pending a review by your lawyer. Many employment contracts have space for changes, and if not, an addendum can be attached to fully protect your interests.

Many large corporations use a form contract that is equitable and will agree to reasonable modifications. However, they will not usually agree to a change in principle or grant you title or a royalty on any inventions that you may make as a result of your job.

An assignment of inventions made in the normal course of your work or as a result of what you learned from the company is almost always required. However, some contracts go farther and require an assignment of any invention that relates to the company's business. Read your contract carefully, so that you understand your obligations.

Treat your company fairly and assign inventions that result from your employment to your company. However, during your contract negotiations, try to exclude any invention previously made as well as those inventions unrelated to your work.

Compare a proposed contract with the agreement of your former employer. A few contracts require an assignment of inventions made within 1 or 2 years of termination. In general, these contracts are designed to prevent an individual from suppressing an invention and filing a patent application in his own name after he leaves the company. Questions concerning your contract with a former employer should be discussed with a lawyer at the new company. In this case, you both have a similar interest in avoiding future legal problems.

Some years ago, Steven Henry accepted a position as a supervisory engineer. Steve had worked for a small company during the previous year, but before that, he had worked for 5 years for the new company's major competitor. To avoid any problem, Steve discussed his earlier contract with a lawyer at the new company.

One provision of the earlier contract required that any company information removed from the premises should be returned to the company upon termination.

Steve had inadvertently taken his personal files, which included copies of his letters, memos, and reports on various projects. The lawyer at the new company called the competitor's lawyer, explained the situation, and returned the unread files to them.

10.2 INTEGRITY

"There is nothing more important than integrity," according to Paul Marinaccio, director of AMF's Research Center.

Dishonest people may get promoted; however, dishonesty is more frequently a basis for dismissal.

Jim Edwards is a talented and very bright scientist. He had received a number of early promotions and appeared to be a promising candidate for vice president of R&D. Jim suppressed negative results and promoted his projects with zeal. As a result, management lost confidence in his organization and Jim was forced to find another job.

At times, you must give management bad news. Trying to suppress it will make matters worse. Remember, engineers don't work miracles. They apply sound engineering principles and report facts.

You integrity is absolutely essential, and at times, should be tempered with diplomacy. For example, Edward C. is a fine engineer who was asked to solve a problem on a major project. After several hours of calculation, Ed realized that there was a serious design error. He told the manager it was impossible to remedy the problem and that the entire project should be abandoned. Ed was correct, but he was also fired. The project was subsequently abandoned, several million dollars were written off, and the project manager was also fired.

Ed should have approached the project manager with his calculations and explained there was a serious problem that could endanger the entire project. He could then explain his initial calculations and inquire about possible alternatives or suggestions.

Ed may still have been fired, but probably not. By reporting facts instead of a conclusion, others may have have become involved. A careful review would have illustrated the seriousness of the problem and could have saved hundreds of thousands of dollars for his company.

Your integrity also involves doing a full day's work for a day's pay. Some engineers think they are hired for what they know, or that thinking should not be subjected to an 8 hour day. By putting forth your best efforts during the regular work day, you will accomplish far more than your peers.

No one is expected to work a 10 hour day on a regular basis. Nevertheless, if a production machine is running scrap, don't leave when the whistle blows.

Many engineers work diligently, utilize their time, and also apply their creative ability to company problems during off hours. They have the highest integrity and the greatest potential for advancement. You can join this group and share the larger raises and important promotions.

When you start to work, don't waste time or get caught up in processing papers. Make an honest effort to direct your efforts to the most important projects. Organize and apply yourself to the task at hand.

Use holidays and vacations intelligently. Recharge your batteries and return to work with a fresh outlook.

In establishing your reputation for integrity, don't make promises you can't keep. Complete your jobs within the promised time period. However, if extenuating circumstances develop, tell your boss and anyone else who is counting on your performance.

There is a story about a U.S. manager at a Mexican plant who demanded a 20% increase in production within 3 months. A U.S. engineer explained that the equipment could not operate at that speed; however, his Mexican counterpart said that there would be no problem. Outside of the manager's office, the American said, "You know that isn't possible." "But of course," said the Mexican, "but that wasn't the answer the manager wanted." It is usually far better to face the immediate wrath of a supervisor than to permit him to promise what cannot be performed.

John J. is a young engineer who told his supervisor that he had obtained the drawing of a competitor's latest development. He claimed he had inadvertently picked it up at a customer's plant. An initial investigation disclosed that his relative worked for the competitor and it appeared that John's initial effort had been to sell the plans to his employer. The company lawyer notified the competitor and advised the employee to return the drawings. He was also advised that under no circumstances should he disclose their contents to any other employee or use the information in his work. John had destroyed his reputation for integrity.

10.3   OBLIGATIONS TO DISCLOSE AND ASSIGN

Virtually every employment contract requires an employee to disclose inventions arising out of his work for the company. Therefore, if you have a good idea and fail to call the invention to the attention of your patent department, you have breached your agreement.

Don't listen to an engineer who tells you that submitting inventions won't do you any good or be discouraged if your company doesn't make cash awards for inventions. Submitting a good idea can be your key to success.

Some engineers think they can get even with a company for some imagined wrong. They decide to keep their ideas secret, leave the company, and use their invention against the company. If you start to think like that, quit your job immediately.

Even a technical drone should be ashamed to admit that he has not had a single good idea in the past year. You, as a creative individual, should have many, and if you want to succeed, will submit them for patent consideration.

As you think creatively and learn more about patents, you will find it easier to identify good ideas and to distinguish your ideas from the earlier patents. Establish a goal to submit at least one idea per quarter. Later, try to submit a patentable invention each quarter. By doing this, you can meet your obligations and establish a reputation as a creative engineer.

10.4 CONFIDENTIALITY

Trade secrets, like patents, enable your company to maintain a competitive edge in the marketplace. The problem with trade secrets is that many engineers and executives do not understand how to establish and maintain them.

Government contractors are frequently forced to follow stringent procedures to prevent the disclosure of classified information. Similar procedures are required to maintain a trade secret.

Many of your peers and supervisors only pay lip service to these obligations. Some will submit proposed papers for legal clearance and, without thinking, will discuss the details of a key process with an outside salesman.

Your obligation to keep information confidential goes beyond nondisclosure. Ted Welton, a group vice president at the Carborundum Co., was responsible for developing a new process to produce granular activated carbon. He worked with counsel to develop the appropriate protection. Each employee (professional and nonprofessional) signed a confidentiality agreement. Each employee received an explanation of its significance. Vendors were not permitted in the plant, and when a pilot plant was set up, it was put in a separate building and access to that building was limited to those working on the project.

Part of your job is to keep vendors and customers out of areas that contain confidential information. Your peers may think it's not their job. They are wrong, it is everyone's job to protect the company's confidential information. Besides, if you want to get ahead, you will do more than your job.

Don't suggest a plant tour for a professional group. If you are asked about plant tours, point out the possibility of divulging confidential information. Explain that a plant tour might preclude the company's enforcement of their rights, even against a former employee who had stolen valuable trade secrets. If tours are authorized, establish restricted areas and insist that all participants be accompanied at all times by a company employee.

You may ask why a company needs to worry about trade secrets as long as it can obtain patents. In many cases, the patent protection is limited and a potential competitor may avoid infringement by engineering around the claims. It is also true that even if your company is using state of the art technology, that information could be useful to a competitor.

In the competitive business world, competitive analysis can be helpful in gaining market share. Business plans, pricing, and costs are key elements which help a competitor undermine your company's market position. It is difficult to keep secrets in today's mobile society; nevertheless, you are obligated to do your part.

10.5 OBLIGATION TO YOUR PREVIOUS EMPLOYER

Some employee contracts prohibit a departing employee from recruiting or hiring other employees. Even if your contract does not have this type of provision, your knowledge of key individuals, hard workers, and creative engineers should be considered confidential. Don't tell your friends how great the new company is or suggest that they apply for a job. Loyalty to a former employer survives termination and you should not misuse confidential information gained in the course of your employment. Besides, you may want to change jobs in the future and would not want a reputation for pirating fellow employees.

However, your obligation to a previous employer should not prevent you from changing jobs in your quest for promotion. In a mobile society, a willingness to change jobs may be a short cut to an important promotion. Therefore, if you want to change jobs, do it. You may be doing your present employer and your wife a favor by changing jobs, since there is probably nothing more miserable than a creative individual who is frustrated by his job.

Ethical companies do not want to steal trade secrets from their competitors. For example, Ralph Rushmer, an engineering manager at the Carborundum Co., asked me to interview an applicant who was working for a competitor. He asked that I explain our prohibition against any disclosure of the confidential information of his present employer. Ralph's secretary called to cancel the interview, because the prospective employee had tried to sell himself on the basis of disclosing his company's trade secrets.

Ralph recognizes the importance of integrity and knew that if we hired the individual, we may have been subjected to a lawsuit and even enjoined from doing business in a desirable market. Besides, Ralph also assumed that the individual would be willing to sell our secrets in the future.

At times, it may be difficult for you to determine what information is confidential. For example, many supervisors and executives feel that information is confidential even though it is in the public domain. Remember that you are free to use published information, but always exercise caution to avoid confusing internal reports with published information.

In general, the courts are reluctant to enjoin a person from working. Nevertheless, the court did that to one engineer who had been hired because of his knowledge of space suits. In that case, the information was classified, so it was easy to establish the existence of trade secrets.

In another case (*Motorola* v. *Fairchild*), the court refused to enjoin Motorola's key employees from working for a competitor. The court refused because Motorola had published most of is information and conducted numerous plant tours without adequte safeguards. In addition,

Motorola permitted vendors in its plant and failed to protect detailed information about their processes.

10.6   WHAT BELONGS TO THE CORPORATION?

Whether or not your invention belongs to the company is determined by your employment contract. Consider the case of John O., who is employed by a large conglomerate to work on inertial guidance systems. John, an avid water skier, invented a new water ski. It is outside his field of work and did not arise out of any knowledge he gained at the company. Whether or not his invention belongs to the company is determined by his employment contract.

John can ask for a release, but in so doing could reduce his chances for promotion, because some managers believe that your interest in pursuing something the company has rejected shows a lack of loyalty.

It may be true that golfers, sports car racers, and other hobbyists devote more time to their avocations than John would in developing a ski. Nevertheless, he should carefully consider the alternatives before investing his time in a sideline business. Besides, the sports car racer may be passed over for a promotion because of incurring needless risks or for devoting too much time to a hobby.

Peter S., a successful entrepreneur, was one of the world's leading scientists in his field. He sought permission to do consulting work for a small company in a noncompeting field, but was refused. An executive stated that he should devote his total efforts to the company's business.

That same executive refused to approve a new venture proposed by Pete, but was furious when Pete left to pursue it on his own. Peter honored his commitments to his former employer and became a successful businessman.

If you have entrepreneurial talent and a desire to work on your own, you may find more satisfaction and wealth by developing your own business. However, you should recognize the conflicts between a successful corporate career and your own business interests. You should decide what is most important to you.

For those who decide to pursue their own business, don't try to use your employer's facilities. By doing so, you may sacrifice valuable patent rights. For example, an employer may obtain a nonexclusive license to any invention developed using his facilities. Therefore, in starting your own business, be fair to your employer and to yourself.

As long as you accept a salary from your employer, live up to the spirit of your employment contract.

10.7   THE EFFECT OF GOVERNMENT CONTRACTS ON YOUR WORK

You should recognize the importance of your patents and avoid any loss of your patent rights that might result from accepting a government contract.

The General Tire Company accepted government funding for the development of a tread compound and in so doing lost their exclusive rights to a patent on extended rubber. A U.S. Court decided that in view of the government contract, Firestone did not have to pay a royalty to General Tire. A newspaper article stated that royalties would have been about $150 million.

In another case, Walter B., an engineer and project manager, believed he could convince the American Gas Association (AGA) to pay for a research program and that AGA would waive any rights to patents resulting from the research. When AGA refused, he tried to convince management to accept the AGA terms. Management refused and questioned Walt's judgment in giving up their rights to exclusivity for a $90,000 research contract.

10.8   DON'T SIGN ANYTHING WITHOUT YOUR LAWYER'S APPROVAL

If you hope to get a promotion, don't sign agreements unless they have been approved by your lawyer.

Some years ago, two engineers visited Bausch & Lomb to obtain information on an optical system for a government contract. They were told that the information was confidential and would not be disclosed, but because of their persistence, they were offered the information if they agreed to keep it confidential.

The two engineers then signed a statement that the information was the proprietary property of Bausch & Lomb and that it would be held in strict confidence. In effect, they agreed that they and their employer were prohibited from disclosing or using the informaton.

They should have been fired. Their employer withdrew a bid for a government contract and warned his engineers that anyone signing a similar agreement would be summarily dismissed.

11

Living with Confidential Information

11.1   DEALING WITH INDIVIDUALS

Beware of outside inventors; they can destroy your career. However, you can protect yourself and your company, by asking for legal advice before talking to an inventor about his idea. Ask for legal advice even though the inventor has a patent.

Individuals submit ideas to corporations and expect to be paid substantial sums of money for these ideas. The best way for you to handle outside submissions and avoid problems is to refer the individuals to the company's patent department. They will usually require the outside inventor to agree that any payment will be based on his patent rights and that there is no confidential relationship. This agreement will be in writing, signed by the inventor.

By basing payments on patent rights, the company avoids any misunderstanding and pays only on products that are covered by the claims of a patent for the life of the patent. Ask yourself why your company should pay an outside inventor for an unpatentable idea when a competitor can use it for free? Of course, some ideas may warrant a nominal payment if they provide lead time over a competitor. However, few outside inventors contemplate this type of an arrangement.

Suppose, for example, that someone in your company is already working on the same idea. The outsider's idea should be promptly returned by the patent department with the explanation that the company is already working in this area.

Suppose your company had developed a somewhat different approach, or started with an individual's idea and modified it to produce a radically different product. How much should the individual be paid? He may believe you stole his idea and that he is entitled to a substantial sum of money.

Accepting information without a written agreement is an invitation to litigation. A lawsuit may be embarrassing, be expensive, and focus attention on your lack of judgment in accepting the information. The damages might also be considerable if the submitted information was inadvertently disclosed.

11.2   DEALING WITH CUSTOMERS

As you are aware, there are advantages to a close working relationship with your customers. However, this type of relationship frequently requires a legal agreement and the exchange of confidential information.

One company had an excellent working relationship with a customer that was the exclusive supplier of a key element for their secret process. This relationship was placed in jeopardy when one of the supplier's engineers discussed the application with a colleague from a different division without mentioning that it was confidential. The colleague viewed this application as a business opportunity. He designed a system using the same components and sold it to the customer's competitor.

The company lawyers were advised of the problem and, with considerable difficulty, negotiated a license agreement with the customer. The license permitted the company to sell the product and to disclose the technical information. The initial cost was substantial and the agreement called for a 10% royalty, which was unrealistic in view of the expected profit. Nevertheless, this payment was less expensive than the loss of a good customer, the cost of litigation, and damages. To make matters worse, the product was a commercial disaster.

Hence, you can do a better job if you make certain that confidential information is not inadvertently disclosed and that fellow employees know and respect the rights of your customers.

One major glass company insisted that all suppliers enter into "standard" confidential agreements as a prerequisite to any discussions. In my opinion, their proposed agreement was patently unfair. Nevertheless, our suggested compromises were rejected and the glass company threatened to discontinue its business with The Carborundum Co., a major supplier. A senior executive at Carborundum criticized my advice and said that if others had signed, why shouldn't he. Fortunately, the subject was referred to Otto Stach, a fine executive and a gentleman, who was also concerned about the proposed agreement. He set up a meeting with the customer and his lawyer.

After a tour of Carborundum's plant, Otto pointed out that we were prepared to enclose a portion of the plant to comply with their proposed terms. He added that we would need a 3 year commitment to buy a set amount of products to justify this expense.

Otto also described the type of technical assistance we provided and convinced the customer that our information was valuable and should be treated in the same manner as the information disclosed by the glass company. With the agreement of mutual obligation, the glass company's lawyer readily agreed to remove the unfair provisions.

Otto created a better working relationship with a good customer instead of capitulating and sacrificing our company's rights.

11.3 DEALING WITH ACQUISITION CANDIDATES

Sometime in your career you may be called on to assist in the evaluation of an acquisition candidate. Evaluating a company for the purpose of acquiring it may lead to serious problems. Some companies seek to avoid these problems by limiting their evaluation to financial data and human resources.

High technology companies with high potential growth are less susceptible to financial analysis. The problem in trying to acquire a high technology company is complicated when the acquiring company is trying to complement its own capabilities. In such cases, there is often a conflict between buying the technology or continuing with an internal development.

For example, a client company signed a secrecy agreement with an acquisition candidate and elected to terminate their discussions. A year later, they introduced an identical product on the market. They paid a large sum to settle the resulting litigation.

You can minimize your risk in evaluating a potential acquisition candidate by conducting an initial evaluation of nonconfidential technical information. An evaluation of the confidential technical information can then be limited to a few key employees who have been advised of the terms of an agreement. Diligence in keeping engineering notebooks and engineering records are added insurance against misunderstandings.

11.4   THE PROCESS MYTH

Don't be misled by a suggestion that it is better to maintain processes as trade secrets than to seek patent protection. Likewise, don't be misguided by arguments that it is difficult to police a process patent or that an issued patent may tell your competitor what you are doing.

These suggestions are dangerous, because few companies are willing to pay the price to protect their trade secrets. Effective security measures are expensive and require enclosed areas, limited access, locked files, signing in and out procedures, and constant monitoring. Ask any government contractor about the expense of handling classified information.

Besides, one of the country's leading industrial spies scoffs at industrial security. He tells of a client company that was unable to duplicate a successful new candy bar. He obtained the exact formula as well as the details of the process within 2 weeks. He got the information in a bar across the street from the manufacturer by "drinking a few beers with the workers."

In other cases, information is leaked inadvertently. A salesman may comment on the type of material he sells to your company. After a few drinks, he may relate everything he knows about your process.

A few years ago, a client company had licensed a dairy process to two companies. On a Tuesday, they were told by their Minneapolis licensee that a Texas dairy would conduct a trial run of the process on the next day. The licensee also identified a former employee of the other licensee as the dairy's supplier.

On Thursday, the licensee in St. Louis notified my client that their former employee had furnished material to a Texas dairy which had conducted a trial run the previous day. The trial had supposedly been conducted in secret.

11.5   TRADE SECRET LITIGATION

George M. (a trusted engineer with 12 years service) thought he had been passed over because he did not have a college degree. He resigned to work for a competitor and was given a month's pay and escorted to the door.

A few days later, we learned that shortly before giving notice, George had taken a large roll of blueprints. We also learned that George had spent several hours reviewing a highly confidential operations manual during the preceding week.

We filed a lawsuit against George and his new employer. Each side spent several hundred thousand dollars in legal fees and the lawsuit was finally settled. A short time later, both George and the vice president who hired him terminated their employment with the competitor.

You should recognize that trade secret litigation is difficult and expensive. In addition, your company may be forced to disclose confidential business inormation during the lawsuit and incur substantial internal costs. Nevertheless, it is sometimes essential to protect a company's trade secrets and to prevent an employee from violating his employment contract.

11.6   PLANT VISITS

When you visit the manufacturing facilities of your customers and suppliers, you will learn about their capabilities and needs. At the same time, you will add to your own experience.

However, during a plant visit, you may be exposed to confidential information and should clarify any obligations of confidence before visiting the plant. A simple letter stating that no confidential information will be exchanged can be used to avoid a serious problem.

If an exchange of confidential information is contemplated, it can usually be handled with a simple agreement. However, don't wait until you arrive at their plant and are confronted with a legal document to call your lawyer. In general, an agreement of confidentiality that extends for 3 years and requires reasonable business efforts to keep the information confidential is reasonable. However, the agreement should exclude information that you already have in your possession, that is in the public domain, or that is brought to you by a third party. A typical confidential agreement is included in Appendix D. If in doubt, ask your lawyer to prepare an agreement before you leave for the plant visit.

11.7   DEALING WITH VENDORS AND CONSULTANTS

One of the best ways to deal with vendors is to restrict them to conference rooms adjacent the reception area. If you do not have such conference rooms, you should at least see that vendors are not permitted in the plant without an escort. Remember, even though a vendor signs a secrecy agreement, he may inadvertently tell something important to a competitor. His job is to sell his product, and it may be difficult to remember what it is you want held in confidence.

Your job is to control access to your company's information, even though many ranking executives are remiss in this area. You can at times frequently avoid problems in disclosing your company's confidential information by enlisting the help of your company's security officer.

Don't forget that it is a salesman's job to analyze your needs and to determine what his company can do to sell you more products. Don't give a salesman more than he needs. Don't disclose your company's trade secrets or market plans unless someone in higher management wants the information published.

Consultants present a more difficult problem. Most consultants enter into secrecy agreements with their clients and need access to confidential information to do their job. Nevertheless, many of them go to professional meetings, talk about their work, and may try to impress potential clients by discussing your projects. If you repeatedly emphasize the importance of secrecy and limit their access to information they need to know, you will protect your company's rights and may help your career. It is not necessary to become paranoid, but be realistic. If your peers ridicule these efforts, ignore them and do your job.

11.8   SUMMARY

Ask the security officer at your company to explain what is done to protect the company's confidential information. Ask what you can do to help. Don't be a gossip. Avoid bull sessions and don't repeat something until it is official or published. Build your reputation as a doer, not as a talker, and move up to a higher position.

12

Using Your Competitors' Assets Without Violating Their Rights

Competitive intelligence is a critical part of strategic planning and an accepted practice in the industrial world. As a general rule, you should learn as much about your competitors as possible, without compromising your integrity. If you hope to be promoted, you should know what is going on in your field.

12.1   DO NOT TRY TO REINVENT THE WHEEL

There is nothing wrong with copying, as long as your competitor does not hold a valid patent. Even if your competitor has a valid patent, you can still copy his product if you can avoid infringing his patent claims. At times, you may succeed by copying a patented product. For example, one senior executive explained that we were losing market share because of a competitor's patented product. We evaluated our defense, and the cost of litigation and copied their successful new product. After several years of litigation, the parties agreed to settle the suit and we obtained a license. The costs were far less than the potential losses.

Some company lawyers are too conservative and will try to avoid litigation at all costs. They fail to recognize that the high cost of litigation is frequently less than the cost of developing a new product. Besides, the risk of losing in court is usually less than the risk of failure in launching an untried product.

When you copy a competitor's product, don't produce a slavish copy. Apply good engineering analysis and come out with a better product. It will help in the marketplace and may avoid a poor image in subsequent litigation. If you design a better product, it may be patentable and could be the key to avoiding litigation.

The U.S. Supreme Court held that a company had every right to copy a competitor's design and sell the products. It quoted an earlier decision: "Sharing in the good will of an article unprotected by patent or trademark is the exercise of a right possessed by all—and in the free exercise of which the consuming public is deeply interested."*

*Sears v. Stiffel, 140 USPQ, page 528.

12.2   CORPORATE INTELLIGENCE

You can find a lot of competitive information, most of it available to the public. It is usually like a jigsaw puzzle—a matter of picking up bits and pieces, sorting them, and putting them together. Your job is to gather and analyze the information. It's something that many of your peers overlook.

Successful managers, like Ted Welton, learned to amplify their efforts by utilizing a company's resources. Follow Ted's example and see what your company is doing to compile information on competitors and competitors' products. Does your library subscribe to a clipping service? Does your company have a systematic approach to monitoring a competitor's activities? If not, find out why and initiate an activity for competitive awareness.

It is almost always easier to promote your project if it fills a need in a legislated market. Your company's Washington office will frequently obtain information on pending legislation that you can use to sell your projects. It can also obtain information under the Freedom of Information Act (FOIA). If the public affairs department can't help, ask your lawyer to obtain information under the FOIA.

Your public affairs department can also advise you of congressional bills that might affect your company or your project. Ask for help and advise them of your interests. What Congress does will probably affect your company, its competition, and your job Work with public affairs and ask for their help in getting appointed to a committee on a science related project.

Don't hide at election time. Make a contribution to the candidate of your choice. Attend one or two fund raisers and get to know your candidates. When you travel to Washington, schedule a meeting with your congressman. You can give him the support he needs to vote intelligently. However, don't forget that most congressmen are extremely busy and will appreciate your concise presentation with factual support.

If you don't know what is going on in your field or what your competitors have published, you are not doing a good job. Many of your executives read extensively and may have read about a competitor's activities. If you haven't done your homework, you could appear naive in promoting your projects.

You can also help your career by forwarding articles on a competitor's activities to your superiors with a short note on how these activities could affect your company. If possible, show how your project will counteract the competitor's activities.

Get to know the people in public affairs (relations) and the importance of their job. Usually, they know how to write and use a style that is acceptable to top management. Solicit their help in editing a report or a proposed article. The real benefit is becoming friends with good people and in working together for a common cause.

In making a presentation to management, comment on the help you received from the Washington office or from public affairs. You may get credit for using the company's resources. Besides, you should support the organization that helps you in your quest for advancement.

A department that specializes in government contracts has useful information on services and products that are sought by the United States government. You should not necessarily seek government funding for a research project, because such funding usually requires the loss of your rights to exclusivity. Nevertheless, knowledge about government interests and the possibility of selling products to the government can help you to sell your project.

The U.S. Government may be a large potential customer. However, after dealing with the government you will realize that Attila the Hun was really a nice guy.

Learn what companies have been awarded government contracts. If a competitor is working on a government research contract, you can monitor his work by obtaining copies of progress reports. Information about past government procurement practices and future government needs can also help you in bidding on government contracts.

After analyzing a competitor's reports, pass the information and your conclusions on to your superiors and anyone else who might be interested in this information. Joe Girard (the world's greatest auto salesman) and other top flight salesmen stress the importance of keeping their names before prospective customers as often as possible.

Sales to the U.S. government can help you in achieving an early payback on an investment in your project. Sales to the government can also help your company produce commercial products at a lower marginal cost and result in a competitive advantage in the market.

If your market information shows that a competitor has an advantage, it may be time to kill your project. A few years ago, Peter Harrington, director of development at the Carborundum Company, analyzed the market for treating industrial waste. He detected a loss of momentum in the expected legislation, reported this fact to management and terminated his program. Pete was then assigned to a new and better project.

One of Pete's predecessors, John K., had more persistence than any engineer I've ever known. John was very effective in using government information in selling his projects and in obtaining government funding when management was reluctant to support his project. He progressed rapidly, but suppressed any negative information. His lack of integrity precluded further promotion.

12.3 FOREIGN SOURCES

You can preview your competitor's U.S. patents before the patents issue, if the competitors file European patent applications. At your request, your patent attorney can ask his European associate to send you copies of the competitor's applications.

This means that you can find out what your competitor is trying to patent, even though the U.S. Patent and Trademark Office is maintaining the information as confidential. The costs are nominal and yet the information may be vitally important to you and your company.

You can also ask the European associate to send you copies of all patents issued in a particular field. This permits you to identify a potential competitor and to notify the appropriate people in your company.

You can use this knowledge of patents to become a more effective manager. You should also use your knowledge of patents to help your company become a more effective competitor in the market place.

12.4 ASSIGNMENT RECORDS

You should review the *Official Gazette* of the U.S. Patent and Trademark Office to learn about U.S. Patents issued each week. A representative drawing and claim of each patent is included. A quick review of the most pertinent classes and subclasses in the *Gazette* will keep you up to date on the patent activities in your field.

Your patent attorney can also initiate a watch service of all U.S. patents issued in a field or assigned to a company.

You can also find older patents by conducting an assignment search for all patents assigned to a competitor. A review of these patents will give you an indication of their trends in R&D or allow you to keep track of the work of a key scientist.

Your knowledge about a competitor may also help you to move ahead of your peers. In essence, your knowledge is the result of persistence.

James P., a young, successful criminal lawyer, had an incredible record as a defense lawyer. Time after time, his clients were found not guilty. His record was the result of meticulous preparation. Your extra effort will help your company and may enable you to climb higher on the corporate ladder.

12.5   THE IMPORTANCE OF COMMUNICATIONS

Knowledge about a competitor will probably not do you any good, unless it is communicated to upper management. You should also pass on competitive information to your friends in marketing, so they can watch for a new product or ask their customers about an evaluation of the new product. Be sure to notify your friends in strategic planning. By becoming a part of a team, you will do a better job and go farther in the organization.

When you read professional journals and see a significant article, pass it on to anyone who can benefit. A short note will help you to receive the recognition that you deserve or a promotion.

Don't forget that you can use personal information, such as engagement or wedding notices, to win recognition as a considerate person. Most parents appreciate an extra copy of an article about one of their children.

In large companies, it may be difficult to obtain recognition. However, don't develop a poor image by arriving late and leaving early or by communicating with poor English or memos that include a plethora of misspelled words.

12.6 RESEARCH AND DEVELOPMENT LIBRARIES

Are you taking advantage of your company's technical library and using its staff? This staff should conduct literature searches, obtain copies of your competitors' financial reports (SEC l0K), and assist you in locating competitive information.

Your library staff may circulate abstracts of technical publications. If so, scan these publications for technical information to help you do a better job.

12.7 LOCAL NEWSPAPERS AND COURT RECORDS

Dick Olson is president of Carborundum's Resistant Material Company because he did a better job than his competitors. For example, Dick subscribed to the local papers in towns where his competitors had major plants. Information about a plant fire or promotions aided him in his strategic planning.

Few people recognize litigation as another source of competitive information. Information on advertising budgets, share of the market, annual sales, pricing strategy, and quotes can frequently be found in a court transcript. Copies of these transcripts are available to the public. Therefore, if you learn that a competitor is involved in litigation, ask your patent attorney to obtain information about the litigation and to keep you advised of its progress.

12.8 PROFESSIONAL MEETINGS

"Attendance at professional meetings is essential for career advancement and for professional growth," according to Dean Batha, director of development at Fiber Materials International. However, if you want to take advantage of professional meetings, you should plan ahead and obtain the necessary approval to attend. You should also indicate the programs you want to attend in the coming year as part of your management by objectives program.

Meeting people can enhance your career if you concentrate on the leaders. Avoid those who are negative about their companies except to obtain competitive information. Above all, avoid having too much to drink. "The most dangerous man in a corporation is a salesman with a martini in his hand," according to a leading trial lawyer. A drunk engineer can probably do the same amount of damage. In all probability, he will destroy his own career. Drink a nonalcoholic beverage, stay alert, and keep you ears open and your mouth shut, except to ask intelligent questions.

One company lost its foreign patent rights because an engineer, in questioning a speaker, referred to their development. The comment was recorded in the minutes and published before a patent application was filed.

At times, your company may want to use professional meetings to disclose competitive information. For example, your company might want to disclose plans to increase plant capacity to discourage a competitor from doing the same thing.

Never discuss pricing with a competitor. Any discussions about manufacturing costs are equally dangerous. Nevertheless, if an engineer is dumb enough and discusses his company's process, you may, by reverse engineering, estimate their cost. If you do so, document the source of your information and the basis for your estimates.

The use of old friends as an information source can be dangerous. It is probably OK to meet with friends who work for a competitor. However, care must be exercised in answering even casual questions, such as, "How is my old project going?" It is better to avoid all discussions about your company.

12.9 DON'T BE SURREPTITIOUS

There is nothing wrong with asking a competitor a direct question at a public meeting. If he knows you are a competitor, he may not answer. One executive asked a competitor if he was going to manufacture a product. The way in which the competitor refused to answer confirmed that executive's suspicion.

In another case, an employee of company A called a former classmate at company B to inquire about company B's plans to enter a new field. The executive at B asked if the engineer was interested in a job, confirmed his company's decision to enter the field, and discussed its proposed schedule. This conversation came to light during subsequent litigation and was embarrassing to both individuals.

Tony S. is a bright, energetic engineer who became involved in the same litigation. He had been transferred recently to sales and was working in a town where company B was building a new plant. Because of his technical expertise, he was asked to check on their progress. He took photos of the plant but committed a technical trespass by walking on their property and was named a defendant in a lawsuit alleging industrial espionage.

There was nothing wrong with taking photos from public property. Aerial photography, on the other hand, is illegal. It is also improper to gain access to a plant under false pretenses.

If you drive by a competitor's plant, record exactly what you did and what you saw. Do not suggest anything illegal, even in jest.

12.10   PLANT VISITS AND LEGAL ADVICE

In Chapter 11, you read about problems associated with plant visits and confidential agreements.

Public tours of a plant, arranged by a professional organization are proper and may be a source of information. As long as you are open in identifying your company affiliation and do not sign a confidential agreement, you can observe their operations and ask questions. Nevertheless, when you write your report of the visit, indicate the circumstances, so you are not implicated in some alleged wrongdoing.

If you receive information in confidence, do not disclose it and if in doubt, seek legal advice.

Before proceeding to the next chapter, reflect for a moment on the importance of doing your job. It is true that you cannot devote all of your time to industrial surveillance and should not become preoccupied with a competitors' activities. It is also true that most of us should do more in this area and that these activities in moderation will help you to do a better job for your company.

13

An Alternative
Source of Income
for Added Recognition

You can win recognition, bonuses, larger raises, and promotions even though you work in R&D.

"Cyanamid presents salesmen with an annual 'Golden Oval Award' consisting of money and paid vacations. In contrast to this, Cyanamid's Scientific Achievement Award, whose winners are treated to a dinner and a limousine ride."*

One reason for the difference in the way companies treat their engineers and their salesmen is that engineers are too far removed from profits. If you want to enjoy larger raises and bonuses, learn to think more like a salesman.

What can you do to increase the likelihood of a larger bonus? You can take a market oriented approach and promote all of the potential applications for your ideas. Be enthusiastic but realistic. Segregate applications that are appropriate for your company from those that are not. Sell your management on the commercial potential of your ideas.

Consider the investment required to bring each application for your invention to the market. If the investment is too great for your company, suggest a joint venture. Also, consider licensing the various applications of your invention as a source of additional income. Don't overlook the fact that many companies are opposed to licensing their technology when the effect is to sell a share of their market. For example, you won't help your career by suggesting a license to a competitor with a 5% royalty when your company makes a 50% gross margin by manufacturing and selling the product.

Nevertheless, companies like individuals, have strengths and weaknesses. Those that are highly successful in marketing industrial products are frequently unsuccessful in marketing consumer products. Therefore, by licensing applications that are outside of your company's major field, you can generate income, offset some of your development costs, and earn management's attention. Licensing those applications for a 5% royalty, with an initial payment and reasonable minimum royalties in areas where your company lacks a suitable sales force, is appropriate and can lead to larger raises, provided you sell your ideas and yourself.

For example, if you develop a new plastic for engineered parts, consider other applications. If it is flame resistant, can it be fiberized and sold as a

---

Chemical Week, page 56. Reprinted from the March 24, 1982 issue of *Chemical Week* by special permission. Copyright 1982 by McGraw-Hill Inc., New York.

flame retardent textile? One engineer followed this course and convinced his employer to enter into a joint venture with a textile manufacturer.

You should recognize that whenever you sell a product, you will have to convince others that your idea can produce a greater than average return on net assets. In other words, you will have to show an investment in your project, with its multiple applications, will produce total revenues far greater than an investment in a single application.

13.1 PATENT LICENSES

When you sell one of your ideas, direct your primary efforts toward company products, but consider the licensing alternative as a way to generate additional income.

Ed Fronko at General Electric is an astute executive with a successful record in selling technology. He sells developments that do not fit into G.E.'s commercial plans. Ed has lots of projects to work with, since G.E. is not usually interested in the smaller markets, yet has many developments that appeal to companies serving those markets. Ed has a professional staff that screens the developments and offers selected opportunities to interested companies. Nevertheless, it is a difficult job and many good ideas are never licensed.

A number of other companies have tried to generate income from developments that do not meet their criteria for commercialization. Many failed because their efforts were undertaken by technically trained individuals with insufficient concern for marketing. Other companies have tried to sell undeveloped ideas, which didn't work in one field, with the hope that they might work in another.

A more successful approach is for you to sell your successful developments for use in other fields. Your can do this by using field of use restrictions. In a patent license, these restrictions will permit you to reserve an exclusive right in one field and, at the same time, offer an exclusive right in another to some other company.

"Licensing is frequently a viable alternative to selling in a foreign market," according to Jack Mummert a retired group vice president at AMF. Jack recognized foreign licensing as a source of income, as opposed to the cost of creating a foreign sales force. You can improve your chance for a larger raise by including this concept in your analysis for filing foreign patent applications.

Don't overlook the Soviet Union and Eastern Europe as prospective purchasers of your technology. The Eastern bloc countries are interested in advanced technology and may provide an exciting market for your invention. Nevertheless, recognize that it is difficult to penetrate this market. To start, do not try to sell undeveloped technology in Eastern Europe. They too have a vast number of scientists and a vast reserve of undeveloped technology. The Soviets, like most American companies, lack the resources to commercialize all of their technology. However, they are willing and able to buy technology that has been proven on a commercial scale.

You might ask how an engineer can set up a licensing program for Eastern Europe. It is really very easy if you use your company's resources. Talk to someone in your company's international department and get them enthused about your idea and its potential. Sell them on the commercial potential for your idea and solicit their aid in selling management on its potential.

You should also consider the possibility of buying technology from the Eastern bloc countries. Consider, for example, the success of the soft contact lens, which was developed in Eastern Europe and commercialized in this country by Bausch & Lomb. Scan Soviet technical articles and suggest potential markets for their technology to your superior.

When you consider the licensing alternative, ask yourself about the benefits of an exclusive versus a nonexclusive license. If an application requires additional development, a prospective licensee may insist on an exclusive license in order to obtain a reasonable return on his investment. Discuss the alternatives with your patent lawyer and consider granting a license with an exclusive period of 3 to 5 years. This approach may allow the licensee to earn a reasonable return on his investment and, at the same time, allow your company to maximize its royalty income.

In granting a license, it is advantageous to obtain a license grant back to use any improvements that are made by the licensee. The grant back usually takes the form of a nonexclusive royalty-free license and is another incentive for a licensing program.

As you are probably aware, it is illegal to use patent licensing to fix prices. You should also use caution in proposing field of use restrictions, because it might be construed as an improper division of markets. You should also know that a requirement that a licensee purchase unpatentable products from your company is also illegal. These and other points can be clarified by talking with your patent lawyer.

Label licensing should also be considered as a unique approach for selling your idea. With a label license, the purchaser of your company's goods receives a license under your patent. This approach can be used as a marketing tool. Nevertheless, counsel should review the program and advise you on the proper language. You can use label licensing to promote your ideas and to earn a promotion.

13.2   BUYING TECHNOLOGY

Buying technology can produce significant income from a modest investment. Remember that when someone else has already developed a product, it seldom pays to reinvent the wheel.

For example, as previously suggested, the Soviet Union and a number of foreign companies are frequently willing to license their technology. Universities also license technology. However, university inventions usually require further development effort. Nevertheless, these opportunities should be considered.

Dr. Dean Batha has an excellent rapport with academia, knows the leading scientists in his fields, and utilizes university personnel in his work. Dean appreciates creative individuals and uses their help to do a better job. He says, "It is desirable to buy technology and minimize the risk of commercial failure."

Jack Mummert is almost always receptive to new technology. Jack recognizes that outsiders also have good ideas and is willing to pay for patent protection. He also understands that all companies have limited resources and evaluates ideas on the basis of sales potential and patent protection. If it fits into his operation, Jack is willing to accept the challenge to make it a successful product.

"Buying technology can get you to the marketplace more rapidly," according to Bruno Miccioli, former director of market development at the Lead Industries Association. Besides avoiding development costs, this approach can give your company a share of an established market. Bruno emphasizes the importance of being first in the marketplace, because he is acutely aware of how much it costs to delay a product introduction.

"Don't speculate in purchasing new technology," according to Bruno, "if it doesn't fit your company's needs, ignore it." In general, try to avoid any initial payment for unproven technology. Until something is established in the marketplace, it is a high risk investment, and a share of the profit is a fair reward for an outside inventor.

13.3 REASONABLE ROYALTIES

When you suggest selling one of your ideas to another company, don't think in terms of an unreasonable royalty. Consider, for example, that a 5% royalty in the United States is usually an upper limit. Another approach frequently used for determining royalties is to calculate 20 or 25% of the savings attributed to your idea as a reasonable basis for a royalty. If you ask for more, you will eliminate the licensee's incentive to make your idea a success. High volume or low margin products usually support a much lower royalty. For example, in the paint or automotive industry, a 1% royalty may be prohibitive.

When you are considering royalties, you will want to give the licensee an incentive to sell the product. If you look at most companies' net income as a percentage of sales, you will see why a 10% royalty is unrealistic. When you suggest an unrealistic royalty, you provide the prospective licensee with an incentive to find a way to avoid your patent.

A royalty based on throughput is almost always difficult to sell. Nevertheless, the Xerox Corporation convinced its customers to pay a royalty on each copy produced on their copier and became a profitable company. This requires a unique product and a strong patent. Remember that the concept is very desirable, but is also very difficult to sell.

Your associates in marketing, your lawyer, and a friend in finance can help you to establish a fair royalty. However, don't forget that you will have to sell a prospective licensee on your proposal and show him that he can profit from using your technology. If you can convince the prospective licensee that you are only asking for a reasonable share of the profits, you can sell him on your proposal.

13.4 NEGOTIATIONS

Mr. Ashtakov of Licensentorg, a Soviet foreign trade organization, is one of the finest negotiators I have ever met. He is well-prepared, focuses on key issues, and has an ability to come up with creative solutions. He is also patient and sensitive to the needs of his opponents.

In general, you will be more successful in negotiations if you have a team effort. You will need a small group consisting of a businessman, a lawyer, and a technical expert. The leader should have both good judgment and the authority to make a deal. Management may establish limits and at times it may be advantageous to break off negotiations and review your progress with top management.

Good negotiators are objective, since it is difficult to negotiate favorable terms if you are personally committed. One lawyer commented that "many top executives make poor negotiators," since they tend to be impatient. Once they decide on a course of action, they give up too much. It seems incredible, for example, that SOHIO paid nearly $1.8 billion for the Kennecott Copper Company when its market value was closer to $1 billion.

Vincent E. Young is one of the most capable negotiators whom I have known. Vince told me about a problem with business associates who become uncomfortable during negotiations. He complained that their nonverbal communications helped the other side. Nevertheless, Vince usually convinces others to accept terms that are important to his company. He does this because he is very capable and well-prepared.

Engineers, lawyers, and businessmen frequently seek the role of chief negotiator. It is an exciting role, but one requiring lots of experience. Some companies solve the problem by employing men like Vince Young or Felix Klass (recently retired from Celanese). At times, companies refuse to permit interested parties to participate in the negotiations.

If you can be objective and provide technical and business expertise, you may find a satisfying career, or a stepping-stone to higher management, by negotiating technical agreements.

13.5 JOINT VENTURES

A joint venture almost always gets the attention of senior management. In essence, a joint venture is an undertaking by two or more companies to work together on a particular project. They pool resources (such as complementary technical expertise or the technical expertise of one company with the marketing capability of another) or share in the risk of failure with a large investment.

For example, the Carborundum Company developed a flame retardant fiber, yet had no experience in the textile field. It formed a joint venture with two Japanese companies. One company had far more experience in producing an organic raw material, and the other was well-established in the textile field.

In general, you may find it difficult to negotiate a joint venture. It is even more formidable to get three companies to agree, since there are so many variables, including structure and management of the venture company, sharing of future costs and profits, selection of markets and manufacturing location, and so on. However, a checklist of considerations for a joint venture is included in Appendix F. You will appear far more knowledgeable if you review it before discussing a joint venture with your superior.

Working on a joint venture is an invaluable experience. You gain exposure to top management, business experience, and frequently an opportunity to join the venture team with increased responsibility.

13.6   UTILIZING YOUR RESOURCES

When working in licensing and joint ventures, you will develop knowledge in the fields of personnel, marketing, and finance. You may also consult with as many as three types of lawyers. A patent lawyer will prepare a license agreement and review any patent clauses; a general lawyer will draft the articles of incorporation, by-laws, distributor agreements, and management agreements; and a tax professional will point out the tax considerations.

When considering marketing, focus on international distribution, use of distributors, and the partners' sales force. Consider how your invention fits into the marketplace and in what country your invention should be introduced.

Financial arrangements are usually complex. Ask how much funding will be supplied by each party, how much money the venture should borrow, and what countries offer the most attractive financing for new ventures. You should also ask how much stock should be issued and at what value.

When you participate in a joint venture or licensing program, you obtain a unique education and prepare yourself for a higher position. These efforts may be the prelude to a promotion.

You may ask how you can become involved in a joint venture. Begin by selling your project and the advantages that can be obtained by seeking a partner for its commercial exploitation.

14

Positioning in the Marketplace

Tom Butler, vice president of R&D at AMF, identifies a shift to marketing as the first essential element for successful innovation. In other words, if you really want to be successful, concentrate on a market that can be reached in a reasonable period of time. Invest your time in projects that can increase profits within the next few years. In selling a project, find out how your marketing group sells your company and its products. Its approach is probably unique and tailored to fit your company. Use those same techniques and their approach to convince management to support your ideas.

14.1   SELLING A TECHNICAL PRODUCT

Positioning is a widely used strategy in advertising and can be used to promote your product and your career. It refers to creating a position in the mind of a potential customer. Think about creating an image in someone's mind, rather than what you are doing to a product.

Managers, like consumers, are bombarded with a plethora of information. Marketing, production, finance, research, engineering, and probably every other organization within a company has a campaign to obtain corporate funds for its programs. Therefore, apply advertising concepts when you try to sell your projects and eliminate ambiguities by selecting information that will impress top management.

You can also sharpen your image with clear communications. "One problem with researchers is that they are cloistered and talk only among themselves," according to Dr. Dean Batha. Learn to communicate in clear, simple English. Be specific, not vague, and learn to avoid technical jargon.

You can write more clearly and improve oral communications if you read William Zinser's excellent treatise, *On Writing Well.* An evening course in communications can also help, and may do more for your career than anything else. Management wants good communications, yet many engineers are notoriously deficient in this area.

Don't use puffery to promote your program; managers have been deluged. In the words of William Zinser, "Avoid clutter."

Whatever else you do, don't try to sell something unless you believe in it. Confidence should permeate your presentations if you want to be effective. Establish your reputation as a person with a positive attitude, not as someone who dwells on problems.

Robert F. is a fine patent lawyer who did not fit in the corporate environment, because he was negative and dwelled on all of the possibilities of losing a lawsuit. If he had said, "There is a possibility of litigation, but we are confident of our defenses," he would not have discouraged management from entering into profitable fields.

You need consistency in advertising and selling. For example, a single advertisement does not establish a reputation—a political candidate tries to get his name repeated as often as possible. Apply the same philosophy in selling your projects and yourself. Then use progress reports and reviews to advance your ideas.

History shows that the first brand name positioned in the consumer's mind gets twice the long term market share as the number 2 brand and twice again as much as the number 3 brand. In other words, emphasize the advantage of being first when you are trying to promote a new product.

It is also helpful to tie your project to a successful company product, since the successful product has an established place in the minds of most managers.

One way to position yourself is to establish an image of leadership. Participate in community or professional activities and get recognition for what you do. Have the public relations department notify the local paper about your activities. Your company will benefit.

Set out to be the best engineer in your field and the best in the company. You can be number 1 in your company if you get things done right, on schedule, and within budget.

14.2   THE DEFINITION OF A TRADEMARK

You are probably familiar with your company's logo (with an accompanying ® to distinguish its status as a registered trademark) and the importance of a trademark in maintaining a reputation for high quality products.

However, you may not realize that what you do has an impact on the value of your company's trademarks. You can help to create an invaluable mark, or through some inadvertent act, you can destroy whatever value has been established.

On June 27, 1962, the Associated Press reported that after 4 years of litigation, a Federal Court had decided that "thermos" had become a household word. Other trademarks, like aspirin and cellophane, had previously been lost, but why thermos?

The American Thermos Products Company had promoted its house mark for over 50 years. It had sent protest letters to anyone who misused the mark by failing to capitalize the "T" or by using it as a noun. The problem was that American Thermos started too late and had allowed thermos to become the name of a product, as distinguished from a brand name.

You should know that a trademark is a word or symbol used to identify a manufacturer's products and to distinguish these products from those of other companies. In other words, a trademark is an indication of origin.

Your customers look upon trademarks as a way to identify your products and an assurance that they will have the same standard of quality as other products sold under the same trademark. For example, many of you probably select Heinz ketchup based on an expectaton of high quality.

Trademarks provide a competitive advantage because they stand for a unique product that is not obtainable elsewhere. One court held that Borden had improperly dominated the lemon extract business by using their "Real-Lemon" mark to drive their competition out of business.

A trademark, if properly used, can last forever and will continue to increase in value as long as it is used on high quality products.

Do not confuse a trademark with a trade name. At times, a trademark and a trade name may be identical. For example, AMF® is a valuable trademark of AMF Incorporated. It is used as a trademark on the products manufactured and sold by that company. AMF is also used as the company's trade name, but only when it is clearly defined as the company.

14.3 SELECTING A TRADEMARK

Your knowledge of trademarks can enhance your image as an astute businessman and can help you gain the respect and support of the marketing organization. Using that knowledge to select a good mark can also help your project to reach the commercial stage. An early mark selection can position your project in management's mind and identify it as a commercial product. Some years ago, Bruno Miccioli, a project manager for a device for producing virus-free water, was looking for a name for his product. Numerous suggestions involved the word aqua, but many other companies had already used aqua. Bruno suggested Aquella, a contraction of the Italian word for a "pristeen shower," as a trademark. The marketing organization was enthusiastic, legal approval was obtained, and the product was identified as the Aquella™ water purifier.

In selecting your mark, you want people to identify with your products and to remember your name. It is also important to create a favorable image in the customers' minds.

Your lawyer will suggest that you choose an arbitrary word, not a descriptive phrase. An arbitrary mark like KODAK is capable of becoming a strong mark and may be extended to prevent others from using the same mark on nonrelated goods. A strong mark can also be used as the basis for a preliminary injunction, which means that your company can stop a trademark infringer without waiting for a trial on the merits of the case.

Marketing personnel often favor the use of descriptive marks, even though they are not registerable and are usually unenforceable. For example, "Super Steel Belted Radial" was held to be descriptive for tires and was not registerable. In other words, anyone is free to use that term on their steel belted radial tires.

In contrast, a suggestive mark like "London Fog" for raincoats suggests a damp day, is not descriptive of the product, and has become a valuable trademark.

Marketing studies suggest that customers prefer short or simple marks that are easy to read and to remember. If you select this type of mark, it can be used in any size or color. However, care should be exercised in selecting a trademark, so that there are no unfavorable connotations in a foreign language.

In selecting your trademark, get marketing involved early and take advantage of their expertise. Remember that you may be prevented from using a name you want, so select several and ask for legal clearance. A trademark search can usually be conducted and approval obtained in 7 to 10 days.

Don't try to use a trademark that is similar to one belonging to another company. The Goodyear Tire and Rubber Company tried to use "Big Foot" for snow tires and disregarded the rights of the Big O Tire Company. After launching a nationwide advertising program, a court enjoined them from using the mark and awarded Big O $19.6 million. The Court of Appeals upheld the injunction but reduced the damages to $4.6 million.

14.4   PROPER USE OF TRADEMARKS

Your rights in a trademark are based on the use of the mark and not on registration. In fact, to obtain a trademark registration, you must actually use the trademark in interstate commerce. It must be used on the product or on a package containing the product. Even after obtaining a registration, you can lose your rights by failing to use the mark. Remember that your use of a trademark in advertising or public releases is not enough to establish your rights.

Registration of your trademark does put the public on notice and creates a legal presumption of ownership. Registration can also form the basis for obtaining foreign rights.

If you promote a trademark as the name of a product or suggest that it be substituted for the name of a product, you may destroy one of your company's valuable trademarks. The courts interpret any misuse of a trademark by the owner or its employees as particularly damaging. You can avoid this problem if you always use a trademark as an adjective and never as a noun or a verb. In writing technical articles or public releases, don't overlook the opportunity to promote your company's trademarks. However, always make certain that you use trademarks properly.

If you fail to spell a trademark correctly or use it in the plural form, you could destroy the trademark. You should also remember that your initial use of the trademark in an article or public release should incorporate an ® for a registered trademark or a "TM" for one that has not yet been registered.

Remember to always capitalize the initial letter of a trademark, place it in quotes, or type it in capital letters so that it stands out as a trademark. When you make your trademark stand out, it isn't necessary to include the name of the product in every instance. Nevertheless, you should obtain legal approval on any use of the trademark to avoid the possible loss of your valuable mark.

14.5   POLICING YOUR COMPANY'S TRADEMARKS

Part of your job is to look after your company's trademarks. For example, if you come across any misuse of a company's trademark, bring it to the attention of your legal department. You should also call attention to any mark that is similar to one of your company's.

Always be alert for any misuse of your trademarks in internal memorandums. The problem with internal memos is that people become careless in referring to trademarks and before long your company's valuable trademark is lost.

14.6   TRADE DRESS AND SECTION 43

Don't try to see how close you can come to imitating someone else's successful program. General Foods introduced "Country Time" powdered lemonade mix and was very successful in promoting this new product. Borden countered with "County Prize," which might not have been confusingly similar to "Country Time." The problem was that Borden copied General Foods packaging, color scheme, and advertising program. As a result, a court issued an injunction and forced Borden to withdraw its product from the national market.

In general, Section 43 of the trademark law prohibits unfair competition or an attempt to misrepresent your goods as those of another. In recent years, the courts have interpreted this section liberally in an attempt to establish some morals in the marketplace.

14.7.   PUBLIC RELATIONS

Your public relations department is responsible for your company's reputation.

"Our job is to promote a favorable image for the company, establish confidence in the financial community and a reputation for high quality products with the general public," according to Don Rosendale, former vice president of public affairs at AMF.

In many companies, public affairs is responsible for advertising and assures high quality and consistency in all information released to the public.

In addition, public affairs usually includes the corporate grammarians, who are trained in the proper use of the English language. They do not permit the publication of poor English in connection with the company's name, because the misuse of English offends many people and is not indicative of high standards. Therefore, when you write a proposed article or public release, ask for their help. Don't be offended by suggested changes in grammar. Get their help and get ahead.

15

Advanced Concepts for Selling Your Engineering Projects

I discovered that what I'd achieve in my selling career was entirely up to me.*

---

*Hopkins, Tom, *How To Master the Art Of Selling*, page 1. Reprinted with permission of Tom Hopkins International, P.O. Box 1969, Scottsdale, AZ 85252.

15.1   INTRODUCTION

Throughout my 25 years of industrial experience, I have noted that successful executives have one common trait—they sell effectively. They have different techniques for selling;but regardless of the technique, they convince others to act on their advice. At the very least, the men at the top have sold their ability.

By now, you should recognize that what you achieve in your engineering career is almost entirely up to you. In fact what you achieve in your engineering career is dependent upon your ability to sell.

You may think that you are a professional and that you did not spend 4 or more years in engineering school to become a salesman. Don't fool yourself. Everyone is a salesman, but you have an advantage because of your technical knowledge and analytical ability.

Don't tell yourself that you can't sell or that selling requires some natural aptitude. In this chapter, we examine a number of successful techniques used by effective salesmen. Many of these techniques can help you to do a better job and to sell your technical projects. By the end of the chapter, you will realize that selling is a simple craft that you will use over and over again.

Many engineers misunderstand a basic concept in selling. They think of selling in terms of high pressure and of convincing someone to buy something they don't need. Don't make this mistake. Think of selling as helping someone to make a decision that will be mutually beneficial. Think of helping someone to make a decision as a form of leadership.

Let me tell you about an attractive young lady that I met on an airplane. She was in her mid 20's and had just won a trip to Europe. She won the trip in a field dominated by middle aged men. Yet, she stepped into this field and after only 2 years outperformed many of these experienced salesmen. She won her European trip by selling life insurance.

You will probably agree that you don't want to buy life insurance and that it is probably a poor investment. Besides who wants to think about collecting on their own life insurance policy?

You might ask how this young lady became so successful in selling a product that so many people don't want. How was she so successful in this highly competitive field, where so many potential buyers won't even want to talk to a salesperson?

This young sales lady believed in her product. She also recognized that a large segment of the population needed the protection that she offered. In essence, she was helping her customers to make a decision to buy something that they needed.

She told me that selling was really easy, and that a majority of her leads came from customers for homeowner or auto insurance. She knew that young homeowners usually needed a mortgage protection policy for their family. She added, "You never use the term life insurance."

You might ask what you can learn from this attractive young lady, who, without a technical background, launched a successful career at an early age. To begin, this young lady believed in her product, If you don't believe in your project, you will probably not be successful in selling it. She was also sensitive to the needs of her customers. Unless you are sensitive to the needs of your customers, you will not be effective in selling your projects.

You can learn another lesson from this young lady. She told me about the difficulty in closing sales and the need for perseverence. She said that she usually made several attempts to close a sale, and there were times when she had to really work to close.

I noticed that throughout our conversation my young companion seemed very friendly. But when she talked about really going for the close in difficult cases, you could sense her anger and almost see fire in her eyes. She was furious that a young husband would abandon his family and leave them almost destitute, rather than paying a small premium for their protection.

Fortunately, you don't need this approach to hard selling. What you do need is a belief in yourself, a belief in your project, a little sensitivity and a lot of perseverence. You can also benefit from a little of this young lady's drive to succeed.

15.2 PROSPECTING AND PRESELLING

Prospecting is one of the major differences between a successful salesman and an order taker. In other words, a successful salesman is effective in finding new customers.

For example, if you were selling cars, what would you do after you sold a car to your friends and relatives? Would you sit in the showroom and wait for customers, or would you do more? If you didn't do anything else, you wouldn't sell many cars.

You are probably wondering what prospecting has to do with corporate engineering and with selling your projects. You, like any other salesman, have to identify your customer or prospects. Actually, it will help if you think of yourself as a salesman and your boss, his boss, and your company's management, as potential customers.

In the beginning, you have to sell yourself. You have to convince yourself that your project has merit. In doing this, list the advantages offered by your project and analyze any weaknesses. Then analyze your prospects, that is, your boss and his boss.

Let me give you an example of how poor prospecting and a lack of preselling killed a good project. Some years ago, I recognized a lack of patent awareness among the engineers at a large corporation. To overcome this lack, and at the same time improve engineering morale, I proposed a dinner for all of the engineers who had filed a U.S. Patent Application or received a U.S. Patent during the previous year.

I also proposed an award for an outstanding inventor, based on the commercial success of a patented product. The proposals were enthusiastically approved by my boss, a corporate vice president, and by the vice president of R&D. In fact, the project had been approved by the president of the corporation.

There was one serious problem. I had not discussed the proposals with a senior vice president who was the ranking corporate scientist. The senior vice president was surprised when he received an invitation, and cancelled the dinner. With hindsight, the senior vice president could have been presold on the basis that the corporate president and two vice presidents wanted to increase inventor recognition and patent awareness.

Bruno Miccioli was a master at selling engineering programs. Bruno discussed his ideas with his boss and asked for suggestions. He got people involved in the project. After a little further refinement, he and his boss discussed the idea with other executives, elicited their suggestions, and finally suggested a formal presentation. By involving all of the decision makers early, they presold the project, then made a polished presentation and obtained the needed funding.

Prospecting and preselling within a corporation are far easier than in other fields of selling. Nevertheless, don't overlook the importance of identifying your customers and getting them involved before you ask for their approval.

Many of the decision makers ask difficult questions. By involving them early, you can overcome their objections and sell your projects.

15.3   PLANNING FOR YOUR PRESENTATION

Take a market oriented approach when you sell your project. In other words, emphasize what is in it for your customer. Don't forget that many corporate managers are evaluated on their contribution to profits. Therefore, focus your presentation on increased profits, on gain in market share, or on a cost reduction that will not adversely effect quality.

Barry Tarshis a great writer and author of *How to Write Like a Pro*, introduced me to ORSON. He introduced ORSON at a writer's seminar and showed me how to apply it to written communications. ORSON stands for Orient, Reveal, Spell it Out, and Nail it down. You may be surprised, but ORSON works just as well for oral presentations.

For example, suppose you want to sell management on a project to produce a new widget, a project that will require $30,000 and 6 months. Your presentation might go something like the following:

1. *Orient.*   Gentlemen, we have an opportunity to increase our share of the widget market market by over 30% with a RONA (return on net assets) of greater than 50%.

2. *Reveal.*   Jim Brown in sales swears that our estimate is conservative and that he could double his sales, if he only had a slightly stronger widget that sells for no more than a 10% premium. Now, we and only we can produce an all new widget that is 30% lighter and 20% stronger than conventional widgets. In addition, our cost of producing the new widget will be about 20% less than the cost of our present product.

3. *Spell it Out.*   What we have developed is a new composite material that is ideally suited for widgets. Our patent attorney says that we should obtain broad patent protection on this material and that the new widget should also be patentable. All we need is about 6 man months of

engineering time and a mold for the prototype. In other words, for a total expenditure of $30,000 over the next 6 months, we can give you a fully tested prototype. At this time, we will also be ready to modify the manufacturing operation and get into commercial production before the end of the year. Let me give you just a few more details. We will have a prototype within 4 months at a cost of about $20,000. We will also have detailed cost figures for manufacturing the new product. The last 2 months will be used for testing, modifying, and finalizing the design.

4. *Nail it down.*   Gentlemen, we know that our arch competitor, Alpha Corporation, is testing a premium priced reinforced aluminum widget. With your approval, we can beat them to the market with a much better product and with a cost advantage. If you agree, we'll get started immediately.

Some engineers would close their presentation with, "Are there any questions?" or "Thank you." There is nothing wrong with either approach, but why ask for questions. If your management has a question, they will ask without prompting. "Thank you" is at least courteous. However, one expert on speaking advises you to say nothing. As a salesman, he suggests that you look someone in the audience in the eye, preferably the key decision maker and watch for an almost imperceptable nod. If you get the nod, you have your approval.

Based on my own experience, managers usually ask questions, about how long it will take to get into full production or what the cost of revising our production facilities will be. You should be prepared to answer questions of this type and offer more detailed figures as they develop.

15.4   HAVING FUN WHILE YOU SELL YOUR PROJECT

There is one key to effective selling. Having the key builds confidence and really helps you to enjoy selling. The key is preparation. Whether you sell used cars, life insurance, or engineering services, your success is dependent upon your preparation.

Don't try to wing a sales pitch. "Be prepared" is an excellent motto and should be applied even before you start to presell your program. Ask yourself what you want to accomplish and what you can tell your boss that will help you to sell your program. Just take a little time to analyze your proposed program and use a few facts from sales, such as their estimate of the market for an improved widget.

Don't be discouraged by the normal nervousness you feel before a presentation. Nervousness is merely pent up energy which you will convert into enthusiasm for your program. You will learn more about using this nervous energy in Chapter 18.

15.5 TECHNIQUES FOR SELLING ENGINEERING PROJECTS

Advanced concepts for selling engineering projects are based on common sense and a few fundamental techniques commonly used by successful salesmen. Nevertheless, many technically oriented individuals overlook this simple fact.

Before you go any further, let me repeat, you do not have to become a supersalesman to succeed in engineering. However, if you learn to apply common sense and a few fundamentals of selling, you will become much more effective in your profession.

No one knows how many salesmen fail because they don't ask for the order. Recognize that it is far better to ask badly than never to ask at all. Besides, with your educational advantage, it is almost inexcusable to ask badly; yet many engineers do just that.

Let's examine a successful sales effort. In general, a successful effort builds to the close. In other words, your effort builds to the point when you can say with confidence, "All that we need is your approval."

Let's start at the beginning. Have you identified the decision maker in your organization? Who has the authority to approve your project? Do you know what the decision maker wants? Does he want a rapid payback, 20% return on net assets, or what? It is far easier to sell him a project that will give him what he wants than a project that fills your needs.

In most cases, it is easy to describe your project and to show how it will fill the needs of the decision maker. Usually, it is a matter of taking a market oriented approach. You are just presenting facts and showing how those facts support your project. You can also use analogies to show how your project will satisfy the company's needs.

After presenting the facts, help the decision maker make the right decision. In other words, make it easy for him to say yes. For example, you might say, "Jim, with your approval, we'll design a new mold, get preliminary estimates for production equipment, and report back to you in 6 weeks."

Actually, your job is far easier than that of a typical salesman, because you only need a verbal consent and, of course, your salary and overhead have already been approved. On the other hand, a salesman usually requires a signature and an actual expenditure of cash. Stop and think how you feel when a car salesman asks you to sign an order.

In some cases, it is difficult to sell yourself and more subtlety is required. Just remember that you are selling yourself every time you sell one of your projects and every time you do your job well.

Alan is a successful patent lawyer who was hired to be the assistant manager of a large corporate patent department. The problem was that the director of patents was reluctant to delegate any responsibility. He had operated this way for many years and didn't see any reason to change during his last year and a half before retirement. He certainly wasn't interested in training his replacement. Subconsciously, he may have even resented the new man.

Alan recognized that one of the major divisions was receiving inadequate attention. Alan assumed responsibility for that division, organized and prioritized their needs, and did all the work necessary for them to obtain patents. To a large degree, he was doing the work of a senior patent attorney, as distinguished from an assistant manager. Nevertheless, Alan kept his boss fully informed about what he was doing. In essence, he was managing the patent work for a division, rather than assuming responsibility for the corporate department. When the director retired, Alan was named as his replacement. He had sold himself by his self-motivation and by doing a job that needed to be done. Hence, one of the best ways to sell yourself is by doing an excellent job in your present assignment.

15.6   LOGIC AND EMOTION IN PROJECT SELECTION

A number of successful salesmen agree that buying decisions are based on emotion, not on logic. For example, when you bought a new "gas guzzler

8" instead of an "economy 4" to commute 20 miles to work, you didn't base your decision on logic. If you are like many car buyers, you are influenced by personal comfort, style, and self-image. You may be surprised to know that my friends in sales insist that emotion is just as important in industrial sales.

In fact, selecting engineering projects also involves emotion. However, based on my experience, the sale of engineering projects usually involves logic, yet emotion should not be overlooked. For example, if your company purports to be the industrial leader in silicon carbide and you disclose a serious threat to that position, it will be easy to sell a program to protect the company's leadership position.

In an earlier chapter, you read about a marketing vice president at a welding company who rejected every project based on a system. He was committed to manufacturing and selling components. In that case, I doubt that any amount of logic would change his mind.

In general, engineers, scientists, and businessmen are trained to use and rely on logic. However, you should not overlook the emotional impact offered by industrial leadership, successful new products, a dominant market position, automated manufacturing, and the like.

15.7   LEADING QUESTIONS AND CONTROL

Leading questions are not reserved for TV lawyers who use them to force a witness to tell the truth. Actually, some of the best salesmen use leading questions before attempting to close a sale. For example, they know that many customers are predisposed to say "No," and therefore use leading questions to put the customer in the proper frame of mind to say "Yes."

These skillful salesmen bring a customer around to their way of thinking and also keep control of their presentation. You can do the same thing and at the same time can involve your audience. For example, ask, "Isn't it true that high quality is important to our customer?" Then pause, and follow up with a question like, "Don't you agree that we will sell more widgets and make more money if we improve our quality, and at the same time reduce the weight of our widgets and our cost?"

Don't overdo this approach. In general, use leading questions sparingly. But on occasion use leading questions to make a more effective presentation. With practice, you can help youself sell your projects with this technique.

Remember that you want to control the presentation. You want to help your manager make the right decision. In general, don't ask a manager what to do. Instead, make a recommendation that is easy to accept, based on your effective presentation.

15.8   STAGING A CLIMATE FOR SUCCESS

Staging may be another word for preparation, but actually it is a lot more. When you stage a polished presentation, you will be more confident and more enthusiastic. You will probably be more successful.

To begin, prepare a listener oriented presentation with a number of visual displays to illustrate your major points. Flip charts can be effective, inexpensive, and easy to handle. They also get you more involved in your presentation. There is one other advantage to using flip charts. You can sketch an outline of the next chart lightly in pencil on one edge of the chart. This sketch will help you to remember what comes next and to make a smoother presentation.

Samples, that can be passed around are even better than flip charts, because they help you to involve the audience in your presentation. When you involve your audience, you will be more effective. Take advantage of your company's in-house capability for preparing slides or transparancies for an overhead projector. Then, when you have your visuals, practice your presentation. No matter what you do, there is no substitute for a rehearsal. It's even better if you rehearse in the actual setting where you will make your presentation.

Don't forget that what may seem appropriate for the laboratory may not be appropriate for the stage. You might ask why so many engineers prepare carefully, have elaborate visuals, and then make the presentation in a rumpled suit, white socks, and scuffed shoes or sandals. Just think for a moment about a fine play with elaborate costumes and a polished performance. Compare that with the same play during rehearsals, with actors and actresses in jeans with scripts in hand. As a member of the audience, which performance would you prefer?

15.9 OVERCOMING OBJECTION

Almost everyone worries about an unexpected question. They worry about not knowing the answer, or about losing everything, because they failed to anticipate an objection. Don't be intimidated by this needless concern, because there is one universal answer. That answer will serve you well, as long as you don't use it too often.

I learned about the universal answer from Clyde Willian. Clyde was preparing a senior executive for a deposition. He explained very carefully "that if you don't know the answer to a question, say 'I don't know'." There is nothing wrong with this answer. Besides, in a deposition you are interested in facts. No one knows everything or remembers everything. Of course, in a deposition, if you do know the answer, you do give them the facts.

Look at this concept a little closer. If you are surprised by a question, simply reply, "I don't know the answer off the top of my head, but I can get back to you later this afternoon, if that's convenient."

The concept of a universal answer is introduced to overcome any concern about an unexpected question. It is *not* a substitute for preparation.

This may come as a surprise, but many good salesmen look forward to objections. They realize that many customers are reluctant to buy. When those customers voice their objection, the salesman can overcome the objection and make the sale. On the other hand, if a customer doesn't tell you what is bothering him, you don't have a chance to make the sale.

There is a secret for overcoming objections. The secret is adequate preparation, during which you anticipate questions and prepare your answers. The real reward comes when someone asks you a question and you answer with confidence.

15.10 MANAGEMENT PARTICIPATION

Get management actively involved in your presentation.

An auto salesman told me that there is one best way to sell a new car. "Get the customer behind the wheel for a demonstration. Let him drive the car. Let him sense the smell of a new car."

Have you ever noticed that most auto demonstrators are taken from the top of the line and include a lot of optional equipment. A good salesman knows that when a customer has a chance to try some of these options, he will buy them. They also know that when the customer sees himself driving the top of the line, he is more likely to buy it.

Have you shopped for a personal computer? I did, and visited a number of computer stores. Three out of four salesmen sat down at the keyboard, typed in information, and showed me what they could do with their computer. The problem was that I didn't care what they could do, I wanted to know what I could do.

Remember, an actual demonstration is more effective than a dozen flip charts and a movie is better than a picture. For those of you who really want to sell a project, put your management behind the wheel.

15.11   THE KEYS TO SUCCESSFUL CLOSINGS

In the selling profesion, if you don't close, you don't eat. It all boils down to getting the order. In engineering, it means getting the approval to proceed with your project.

Salesmen and engineers have a lot in common. In either case, your success is dependent upon your own efforts to convince someone else to buy your services or product. Your success is dependent upon convincing someone to buy today, instead of putting it off until tomorrow.

Engineering managers are like customers, because they like to postpone decisions. After all, if you can keep your money in your pocket, you may find a better deal tomorrow. If a manager defers an investment decision, he may obtain more facts that will enable him to make a better decision.

Part of your job is to help management make intelligent decisions that will benefit your company. At times, major decisions should be deferred, but it is also true that minor decisions should be made quickly, so that you can get on with your job.

For most engineering projects, you only have to ask management for their approval. Don't ask what to do or what they want. Instead, say something like, "I'm confident that we can complete phase I in 6 weeks at a cost of less than $10,000. If no one objects, we'll get started and have a prototype by the middle of October."

15.12   EFFECTIVE SALES PRESENTATION

In summary, you have the tools for making an effective sales presentation. You know the importance of a market or customer oriented approach, of preparation, and of asking for approval. Just design every word in your presentation to obtain their approval. Every sentence should build toward your success.

For any one who feels uncomfortable in a selling situation, there are two approaches that will help you to become a more dynamic salesman and a better engineer or scientist.

The first is to make friends with one or two good salesmen. Ask them about their selling experiences. Get them to tell you how they closed difficult sales. If possible, work with the company sales department and obtain approval to accompany one of the salesmen on several sales calls. Perhaps you can make a contribution and help your company sell its products.

The second approach is even easier and it's one you can practice almost everyday—observation. Observe retail salespeople. Notice what separates the good ones from the bad. Notice how many greet you with, "May I help you?" Most people answer "No," and keep on saying "no" until they leave the store.

Compare that approach with a salesman in a better store who says, "Good morning, isn't it a beautiful day? Would you like to look around a bit, or do you have something specific in mind?"

One day I observed a salesman in a ski shop talking to a customer about two different bindings. He noted that both bindings were excellent, but that one (slightly more expensive) had an additional safety feature. He also stated, "If you are an intermediate or advanced skier, I would recommend the added safety." He went on to close the sale.

The final approach to improving your effectiveness in selling is practice. You can get this practice by preselling your associates, your boss, and whenever possible his boss.

Remember, when you sell your projects, you are also selling yourself.

16

Strategic Planning and Its Impact on Your Job

Development of new and existing technologies has always been one of AMF's key strategies. *

*Nelson, Merlin E., in *AMF Patents—A Corporate Manual*, page 4.

The challenges for engineers in the 1980's are enormous, but are fairly easy to identify. The difficulty is not to determine what your goals should be, but to determine how to achieve them.

In promoting your ideas, think of yourself as the chairman of the board and chief executive officer of your own venture. You are the marketing, financial, and personnel executive—with one goal, getting promoted.

Companies and managements are unique, yet certain similarities are common to many. Understanding these similarities can help you to achieve your goal.

16.1   THE ENVIRONMENT AND THE LARGE CORPORATION

"Many of our country's chief executive officers are concerned because of a declining rate of growth in today's economy," according to Len Candee, staff vice president of planning at AMF. Consider how you might increase profits in a contracting economy compounded by increased foreign competition and government regulations? For example, think about the impact of Japanese cars, television sets, calculators, cameras, and other foreign products which have had devastating effects on a number of U.S. companies.

A rise in the cost of petroleum based products and a consumer who is frustrated by inflation contribute to your company's problems. In addition, the increase in paperwork required by government agencies and a more litigious society has taken a toll on corporate profits.

The United States has a reputation as a leader in technological innovation. However, our emphasis on short term profits and compliance with government regulations is eroding this role.

For example, one successful division president rejected an allegation that his definition of long range planning was 6 weeks. He added, "Long range planning means next Monday morning."

Nevertheless, today's modern corporations need technical innovations. Your job is to come up with new products and technical solutions to your company's problems. You can sell these products effectively by becoming more familiar with the problems facing your company. Remember, if you don't sell yourself and your projects effectively, you are not really doing your job.

Therefore, if you want to become more successful in today's competitive environment, form an alliance with your top management, convince them to invest in your programs, and perform better than your associates. When you apply the principles of strategic planning and an understanding of management thinking to your ideas, you maximize your probability of success.

16.2   STRATEGIC PLANNING IN A LARGE CORPORATION

Have you considered how your company might solve the complex problem of a declining growth rate, inflation, and increased competition? Have you recognized the need for new business opportunities to make up for the lack of growth attributed to existing products and the increasing importance of competitive analysis?

Len Candee says, "The basic question in strategic planning is how should your company position itself to compete." In other words, you should analyze the present business environment for your industry, your own company, and your competitors before you formulate a strategic plan.

Start by asking what your business is doing now. Do you know where your company is trying to go? If you don't, you cannot do an effective job selling your ideas. Consider the case of Kennecott Copper Co., essentially a one product company. Kennecott had the highest production costs in the industry, according to an estimate in the *Wall Street Journal*, which reported Kennecott's costs for copper at about 90 cents per pound. The market price for copper had fallen to 68 cents per pound and foreign producers continued to sell at this lower price. It seems obvious that Kennecott needed either a massive cost reduction program or a profitable line of new products.

You too can analyze your company's relative position in its present markets and your company's strengths. Can you identify any market trends in your industry? Talk with your friends in marketing and learn about changes in the market. Talk with someone in strategic planning about your analysis and how your ideas can help the company meet its profit goals. What you learn from these sources can help you to sell your ideas.

Your friends in public affairs should have information on pending legislation and industry trends, such as the switch to nonleaded gasoline. For example, public affairs (the competitive analysis group) should also be able to identify any important industry opportunities or threats.

Gather this type of information and use it in promoting your projects. Don't rely on generalizations, use specific examples. Make a systematic approach to obtain information, analyze it, and use it to sell your ideas.

In seeking information on strategic planning, recognize one caveat. Companies usually consider their strategic plans to be highly confidential. In fact, some executives are almost paranoid about the sensitivity of this information. However, much of it is readily available from public releases.

Information about your industry and your competitors is usually attainable. Find out what your competitors are doing and what their perceived strengths and weaknesses are. Also consider which companies are potential competitors, that is, which companies have the technical and marketing capabilities to enter your field. Have you read the annual reports of your company's competitors?

Ask yourself what your company should be doing, in view of your analysis. What alternatives are feasible? How does your project fit into this scenario?

You should look closely at a structural analysis of your industry, and how your company fits into this environment. What are the driving forces of your industry? Have you considered the effects of interest rates, petroleum prices, housing starts, and new car sales on your business. There are many other outside forces which cannot be controlled.

Consider how your industry compares with those of your customers and suppliers. For example, does scarcity of raw materials put your company at a disadvantage with respect to competitors who have their own source of raw materials?

Don't forget to consider the position of your major customers. Does your customer have an incentive to manufacture his own products, instead of buying from your company? One executive who suspected that a major customer might be considering the manufacture of its own ceramic material announced a major plant expansion and reduced prices in view of the economies resulting from a larger operation.

Have you considered what substitutes are available for your products or if the feasibility of using substitutes is determined by price? In the abrasives industry, for example, in many cases there are better abrasives for particular applications, but they are often too expensive.

The definition of an industry is not the same as the definition of where your company wants to compete. In some cases, you may want to avoid certain industry segments because of low margins or because another company has a dominant position in that segment. A dominant position is defined as having the largest market share and one that is larger than your next two competitors combined.

Find out if your company is a cost leader, a cost follower, or a cost cutter. Does your company enjoy a reputation for excellent service, which justifies a higher price? Some years ago, a small welding company had the reputation of charging at least twice the price of a competitor. Nevertheless, its fine engineering staff enabled it to succeed by selling sophisticated installations.

Evaluate your competitors and potential competitors carefully, and try to determine their areas of vulnerability. Analyze their financial strength and any aversion to risk. Inquire about their history in patent and trademark litigation. Have your patent department obtain copies of the decisions. In some cases, it is desirable to obtain copies of the court transcripts as a source of competitive information.

You can also take advantage of professional meetings to talk to former or present employees of your major competitors. Inquire about incentive programs, the competitors' management, and their philosophy. Try to determine if they have a short term preoccupation.

Your competitor's annual report and 10K (filed with the Securities and Exchange Commission) provide hints as to the divisions about to be divested and those receiving larger sums for capital investment. All this information can be used to sell your ideas.

You should also recognize that strategic planning is based on bits and pieces of information gathered from various sources including, but not limited to, press releases, financial reports, customers, suppliers, and professional meetings. You should put as much information together as possible to understand your competitors.

Consider the effect of a recent merger or acquisition. For example, after the Kennecott Copper Co. acquired the Carborundum Co., Kennecott diverted much of Carborundum's cash to its copper operations and almost destroyed the acquired company. The resulting management exodus further weakened the Carborundum company.

Some companies have a strong emotional tie to a particular field. For example, the Norton Co. in Worchester, Massachusetts, has for many years been a leading company in the abrasives field and will fight fiercely to maintain or increase its share of the abrasives market.

In summation, list the strengths and weaknesses of your company, its competitors, and potential competitors. Consider your company's products, distribution system, and marketing capabilities, as compared to those of its competitors and potential competitors.

Don't overlook manufacturing costs, location, technical sophistication, and flexibility. Review recently acquired patents, published foreign patents, and technical periodicals to evaluate a competitor's technical capabilities. Talk with someone in your company's strategic planning group about your evaluation and about potential new products that might adversely affect your company's market position.

Knowledge about your company and its competition will help you to do a better job. Use it wisely and you will be far more successful in selling your projects.

16.3   DECENTRALIZATION AND ENGINEERING

Decentralization in large companies puts more demands on the division general manager. Decentralization also deprives corporate management of the hands on feel for what is happening in the marketplace. You can give management the help it needs and at the same time sell your projects more effectively.

A number of companies categorize their business into stars, cash cows, dogs, and problem children. Cash cows are milked and dogs are sold. The stars receive most of the capital for investment and the problem children are given a great deal of management attention. Demonstrate how your project fills the needs of a particular business and how it can become a star.

Your goal is to capitalize on your ideas. To do this effectively, present your ideas as business opportunities, new products, means of expanding markets, methods of increasing manufacturing efficiency, or other competitive advantages. Show how your ideas fit into your company's business strategy. Recognize that your contribution could convert a dog into a star and gain enough recognition to get you promoted.

16.4 CORPORATE STRATEGY FOR RESEARCH AND DEVELOPMENT

One problem for top management is to provide adequate human and financial resources to businesses that will maximize the long term benefits to the company. As a member of the research and development group, you can help in selecting the most promising projects.

Your role in a research and development group may vary from doing basic research to finding new business opportunities that will complement the company's present products. In either role, you will help management select projects that will maximize the long term benefits for the company.

In selling your project, tailor your approach to the organization. Concentrate on the needs of the market and profit motivation. Reserve the technical intrigue for your fellow engineers.

You should also recognize that a divisional approach to research and development frequently results in a concentration on minor improvements, shop problems, and little or no basic research. On the other hand, a central laboratory usually takes a more academic approach, which is not always in tune with the needs of the market. Find out where you fit in and sell yourself and your projects more effectively.

16.5 TACTICS

You can usually sell a small project more easily than a large one. Therefore, break your large programs down into more manageable projects. Then structure the projects to be done sequentially, to reduce the risks as investments increase.

Be sensitive to the needs and feelings of your audience. Recognize that higher management usually has considerable expertise, but remember that a member of management may not be technically oriented and may be turned off by a technically oriented presentation.

In selling, speak English. For example, refer to red and white blood cells instead of arythracytes and leukocytes. Remember that you and your superiors have corporate profits as a common goal. In many cases, its merely a matter of speaking with confidence and presenting facts to show how your products will give your company a competitive advantage.

16.6   SUMMARY

Above all, apply the concepts of strategic planning to manage your own career. Analyze your company and its philosophy. Do it well and be prepared to take on a broader responsibility in a way that will give management what it wants.

One personnel executive commented that many engineers thought they were ready for broader responsibilities, but were not doing well in their present jobs. Never forget that doing a good job in your present position is an important step in getting promoted.

17

The Art
of Negotiation

17.1 INTRODUCTION

Clyde Willian of Willian, Brinks, Olds, Hofer, Gilson & Leone Ltd. of Chicago, Illinois is one of the best patent litigators in this country. He is also one of the most capable negotiators that I have ever met.

Clyde believes in thorough preparation, excellent communications with his °client, and a team concept. For litigation, Clyde's team typically includes a younger associate, one of the client's lawyers, and a technical expert. In one case, Clyde, John Pavlac, his associate (one of the best legal writters that I know), a technical expert, and I were preparing for a complex trial that was scheduled to begin within a few days. The lawyer for the other side called and suggested one more meeting in an attempt to resolve our differences. We agreed. Then Clyde and I decided that Clyde, John Pavlac, and the technical expert would continue to prepare for trial, while I went with our business unit manager to negotiate a settlement. In this way, we maximized the utilization of our resources and put ourselves in a better bargaining position.

When I went to the negotiation, the trial lawyer for the other side asked if Clyde was coming. I explained that he was continuing to prepare for the trial. We settled the suit on more favorable terms than I would have expected, which our strategy helped us to obtain.

Preparation and strategy are important for intercompany and intra-company negotiations. In other words, preparation and strategy are important for your everyday negotiations. When you use these techniques wisely, you will become far more effective in your job.

One lesson that you can learn from Clyde Willian is that there is no substitute for adaquate preparation. For example, before I left for the negotiation mentioned in the preceding paragraphs, Clyde and I discussed all the conceivable alternatives and reviewed them with our top management.

Whenever you negotiate with the representatives of another company, it is imperative that you keep your management fully advised. I recall another situation when a federal judge scheduled a meeting in his chambers; he also ordered the parties to bring someone with the authority to settle the suit.

Before attending the meeting, I talked with Robert Straka a young aggresive vice president (now deceased), and reviewed our strategy.

During the meeting, the judge suggested a unique approach. I called Bob and recommended that we accept; he agreed. Later, the other side breached this agreement and we obtained a permanent injunction that prohibited them from infringing our patent.

This illustrates that conditions frequently change during a negotiation and at times you may want to ask your management for approval. However, if you are well-prepared and have discussed the alternatives, approval may be a formality.

Now, consider your role in a typical negotiation with your superiors. Assume that your widget project is in trouble. The prototype failed during a test and you need more money for a new prototype. Your boss suggested terminating the project, but you asked for a few days to review the project and then make a recommendation.

If you believe in your project, you will want to convince management to provide the additional funding. However, in any negotiation you may have to make concessions.

To begin, you analyze the results of your previous work, recognize problems, and consider alternatives. For example, you might need to redesign the prototype, expedite other tests, continue on a reduced scale, or ask for substantial sums of money.

Assume that you suggest a joint venture as a means for funding the project and a number of potential partners based on their financial, marketing, and technical capabilities. After reviewing a checklist such as the one shown in Appendix F, you discuss your proposal with management and are authorized to meet with one or more of the potential partners. Sometimes management elects to talk with more than one potential partner to find out who is the best candidate.

When you begin the negotiations, you will probably ask for control of the project. You may find that your proposed partner has far more to contribute than you anticipated. Besides, almost everyone wants to control a joint venture and is usually reluctant to take a minority position.

In such cases, you may find it necessary to share control of the venture and to require approval by both parties for expenditures above an agreed amount. Frequently, one party provides the president of the venture and the other provides the controller.

During your initial meeting, you will make your proposal and answer questions. In some cases, depending on how much your proposed partner knows about the project, you may recieve a counter proposal, although that usually comes later.

There is almost always one really difficult problem in a negotiation of this type. The problem involves the valuation of the contributions made by each party. For example, you point out that you have a basic design for a computer and broad patent protection. You add that your company has already invested well over 1 million dollars in the project. Besides, the early investment involved greater risk than the subsequent contributions. Your proposed partner counters your argument by talking about their advanced manufacturing capability and worldwide marketing organization. They too have a significant investment.

In general, you will have to concede some points, compromise others, and hold the line on those that are most important to your company. In any event, your skill as a negotiator will help your company obtain a favorable deal.

Before going further, there is one note of caution. If you or your company are unreasonable, you will either kill the deal or the project. Just remember, inequitable arrangements seldom lead to success.

17.2 DIPLOMACY

John is one of the rudest individuals that I have ever met. He doesn't win many friends or influence many people, but he uses rudeness as a technique for negotiation. He insults the other side and tells them that they are completely unreasonable for failing to agree to his demands. Usually, the parties reach an impass when they realize that if John is this difficult during the courtship, he will probably be worse after the wedding.

What John fails to recognize is that you don't usually advance your cause by antagonizing the other side. His approach is immature. It is like two drivers that I saw after a minor accident. They had both raised their voices and were loudly accusing the other of some inept or stupid action. All each did was to further alienate the other party. After all, the insurance companies would decide who pays whom on the basis of the unemotional facts.

Some engineers and scientists overlook the importance of the social amenities in negotiations. For example, a German engineering manager started a meeting with the representatives of another company with a direct confrontation. He said, "You are infringing our patents. How soon will you stop and how soon will you pay us for past damages?"

Don't overlook the importance of getting to know the other party. Don't forget that it is far easier to make a deal in a friendly environment than in an antagonistic one. In addition, it is just common courtesy to ask a few questions such as, "Did you have a good trip? Would you like a cup of coffee? Were your accommodations satisfactory?" A statement like, "We appreciate your coming," also helps to start the relationship off on the right foot.

Furthermore, if you fail to settle your differences with another company, you may become involved in litigation. However, a businesslike mutual respect, that is, one without antagonism may lead to a settlement of the lawsuit long before it consumes large sums of money.

Consider diplomacy in an office setting. Joe is working on an important project and needs an additional technician to meet his schedule. He also knows enough to avoid a direct approach such as, "I must have an additional technician, today."

Instead Joe starts by saying, "John (Joe's boss), we have a bit of a problem on the widget project and I would like to talk about our alternatives and get your input. You probably know that my number one technician is spending half of his time on the beta project, which is also on a tight schedule, and that the drafting department has to complete the beta drawings before starting on our widget program. I think that you will also agree that the company wants the widget project completed within the next 6 months."

Notice the technique that Joe uses in seeking an accord on a number of issues. In effect, he is bringing his boss into concurrence with his views and assuring himself that he has an agreement. He is also testing his boss to find out if the time schedules are a real, rather than an imaginary, problem.

Next, Joe suggests several alternatives. For example, "John, there are several solutions to the drafting problem. We could authorize overtime, which would put us a little over budget, or use outside services and stay within budget. I know that there is a corporate mandate to use inside services, but the only other alternative that I can think of is to delay the project and miss the big trade show. I personally would prefer using the outside draftsman and believe that we can keep the costs to a minimum."

Joe then goes on to the second problem. "John, our second problem relates to the assembly of the widget subsystem. Harry, my best technician, can't complete the subassembly for the widget project and the beta project within the next 6 months. We could authorize Saturday work, and probably make it, but I think there is a better solution. If we used the new technician to work with Harry, he could do all of the mechanical work on the subassembly and that would allow Harry to concentrate on the electronics, where he is really good. The new technician could also help on the beta project and we would be able to bring both programs in on schedule. Besides, the new technician is still in his training period, so we will only be charged for half of his time. For this reason, we won't overrun our budget by more than about 5%."

Notice how Joe provides alternatives rather than asking a yes–no question. You may also have noticed the use of a trial balloon to test John's reaction. Joe used "I think" as a way to test the water. When he didn't get an adverse reaction, he proceeded. "John, it would make things a lot easier if *we* could get the new technician by next Monday, but if we can't swing that the following week would probably keep us on schedule. On the other hand, if we don't get some help by then, I think that we had better tell top management that they can't count on having the widget and the beta on schedule. John, do you see any other altervatives?"

Finally, Joe closes his presentation with, "John, would you prefer to talk to personnel about releasing the new technician from his indoctrination program or should I?"

Obviously, Joe thought through his approach in advance and carefully staged his presentation. He approached his boss to seek help in solving a problem and yet he diplomatically suggested solutions. He also invited comments and used an alternative closing technique to nail down the sale.

17.3 SENSITIVITY

Sensitivity is the combination of three components. First, you listen to what the other party has to say. Second, you think about what they say. Third, you consider their needs.

I recall one young attorney who had a serious problem—he failed to listen. Like many others, he seemed to have selective filters that screened out everything that wasn't in agreement with his view. Such people are usually so busy thinking about their arguments that the other person's views don't register in their mind. This young attorney went even further. He was concentrating so much on his argument that he didn't even hear the other party agree with him. Incidently, this is not an uncommon problem.

What happened was that the young attorney and I were discussing a lawsuit with my boss, who was reluctant to follow our recommendation. After a lengthy discussion of the alternatives, our boss said, "Okay, go ahead." The young attorney then presented another argument that caused our boss to reconsider the matter. After another review of the alternatives, he agreed again.

The moral of the story is to listen, and after you have made the sale, shut up. Actually, you don't really shut up, but change the subject or extricate yourself gracefully. For example, after getting an agreement say, "Thanks Joe, we will keep you fully informed and do everything that we can to keep the costs down." Another approach might be to say, "By the way, there is another matter that we need to discuss, are you free sometime this afternoon?" Incidentally, the young attorney did learn to listen and went on to become very successful.

If you want to be successful, you can't put too much emphasis on sensitivity and learning to listen attentively, since negotiation frequently includes the use of subtle indicators. For example, when your adversary changes from, "Our management will not accept that," to "I don't think I can sell that to my management," he may be telling you that he is prepared to give in on that point. If he says that he thinks they may go along with a particular demand if they can have thus and so, it is useless to continue to argue about that point. Yet, many inexperienced negotiators miss or overlook these signals.

Fortunately, there is one technique that can help you to become more sensitive and, at the same time, more effective—repeat your adversary's position. This also gives you additional time to think. Just say, "If I understand you correctly, your view is . . . ." At times, you may even want to reverse this technique and ask the other side to repeat your position. Say something like, "Since we seem to differ on this point, would you restate our position as you understand it?" Or you might say, "Perhaps it may help to clarify the situation if I state my understanding of your view and then you give me your interpretation of our position. Perhaps we will find that we are a little closer than it seems."

In one case, two senior attorneys were discussing a possible settlement and didn't even realize that they were basically in agreement. I listened for awhile and finally interjected a comment to the effect, "It seems to me that you are basically in agreement." I then repeated the elements on which they agreed. The final issue was a matter of semantics.

Another point that many inexperienced negotiators overlook is consideration for the needs of the other party. For example, some years ago, I was a member of a negotiating team that conceded too much too soon. We had failed to anticipate a final negotiating session which was conducted by a key executive. The executive expected to obtain additional concessions. To save the deal, we made more concessions than we had anticipated.

You should always consider the needs of your opponent and try to structure a deal to meet those needs. For example, the other side may want your company to hold them harmless against patent infringment, yet your company does not want to assume the risk of a lawsuit that might cost several hundred thousand dollars. By recognizing their concern about a major investment and a possibility that they could be precluded from selling a product, you might suggest a "hold harmless agreement," wherein you share the costs with your company having a ceiling of two hundred thousand dollars, and provided that the other company accepts your price. At the same time, you should have anticipated this problem and proposed a slightly higher price. Then, of course, if they drop their demand for a hold harmless, you can make a further concession on price.

There is one other aspect of sensitivity that should not be overlooked. Many negotiators use nonverbal communications to signal their acceptance or rejection of a point. I recall one negotiation when I sensed that the other party had accepted a point and proceeded to another. My young associate later asked why I had changed subjects and didn't believe that the other side had agreed, when in fact, they had.

Sometimes, you can pick up an almost imperceptible nod and will not need to press for an affirmative confirmation. At times, it may only be a conditional acceptance, which really means, "You can have this point, but I'll expect some thing in return."

Watch for those imperceptible nods or head shakes as well as a few of the more obvious indicators. For example, if your adversary across the table sits back, folds his arms across his chest and crosses his legs, he is probably rejecting your proposal. You should also watch increased nervousness or discomfort as an indication of a sensitive area.

Above all else, when you participate in a negotiation, don't undermine your team by your own nonverbal communications. All you have to do is be objective and relax.

17.4 COOPERATION

Inequitable deals seldom lead to a successful conclusion. In other words, make certain that there is enough incentive for both parties to make the deal successful.

I recall one major negotiation that resulted in a multimillion dollar joint venture. Bob B., the attorney for the other side, was extremely cooperative and yet did not concede points without extensive negotiation. Bob also gained at least one point for each point that was conceded. Bob believed in the concept that we were really working together to define a fair solution. Bob is an exceptional negotiator and a person who is a pleasure to work with.

If you have any hesitancy about establishing a mutually cooperative arrangement, just remember one thing. Most agreements are renegotiated within a few years. You should also recognize that in intracompany negotiations, you will probably encounter the same individuals in the future and may have to negotiate and renegotiate with them throughout your career. Besides, the last thing that you need is an enemy. If you persist in forcing inequitable arrangements on others, you will create enemies.

Reflect for a moment on intracompany negotiations, whether they involve a larger salary or increased funding for your favorite project. You and the other negotiator are both members of the same team and are both trying to do what is best for your company.

Take salary, for example, and imagine that you believe that you are being underpaid. It is not, as many people assume, advantageous to provide below average compensation. A company that pays less cannot attract or keep above average employees.

Now, let's imagine that you want to discuss your views with your boss and that your goal is to obtain a reasonable raise. After considerable preparation, you might start by saying that you are not satisfied with your present salary and then repeat your understanding that the company's policy is to reward extraordinary performance with above average wages.

You might then mention a number of your accomplishments, and that you have worked many hours of overtime without additional compensation and expect to continue. You might ask if the company really expects to maintain a high level of morale with less than average compensation. However, I suggest that you ask a "Yes" question, rather than one that will lead to a "No" answer.

Don't be misled. Salary negotiations are not usually simple and are more frequently combined with an appraisal of your performance. However, combining a salary discussion with a performance appraisal gives you a further advantage to probe and obtain a candid review. For example, you can ask, "If I do what you ask, will you provide another review in 6 months, with an additional raise based on performance at that time?"

As I recall, there was only one time in my own career when I attempted to negotiate a higher salary without changing jobs. I had received an outstanding performance appraisal and what I considered to be a less than satisfactory raise.

At this time, I had not developed much skill in negotiation, but had gathered sufficient facts to support my case. I asked for an honest appraisal of any deficiencies and what I should do to earn a larger raise. My superior was very evasive and suggested that it was merely a matter of getting more experience.

Several months later, this same man was shocked when I resigned and took a better job. He was surprised many times both before and after my departure, because he failed to recognize the inflationary effect on the wage scale. Unfortunately, my colleagues and I failed to educate him on the importance of staying competitive with other industries in the area.

17.5 PREPARATION

Preparation is so important for negotiation that it deserves a separate section, even though it is mentioned in the previous paragraphs. Another reason is to discuss how you can become more effective in preparing for a negotiation.

All you need to do when preparing for a negotiation is write down each point that may be in contention. Mark those that are most important to you and your company. Then list the reasons why the other side should concede these points. After you do that, list the reasons the other side may have that you should concede each point. Try to analyze the proposed settlement through their eyes and to identify which points are most important to them.

Once you have written down all of your expected points of contention and the arguments pro and con for each, you will want to discuss your proposal with your superior. However, even before that, it is a good idea to rehearse your proposed presentation. Rehearse it two or three times, possibly with your spouse or with a colleague. It would be even better if you found someone to play the devil's advocate.

For intracompany negotiations, you have the advantage of knowing the parties. Nevertheless, your preparation should include an analysis of the other party. Find out what they usually look for, what type of questions they ask, and if they have any idiosyncrasies or negotiating techniques.

In some cases, you can even analyze another company's negotiating team. First, find out who will attend and what their positions are within the company. Check the library to see if any are listed in *Who's Who* or if they have published a technical paper. If they have published an article, you can frequently get things off on the right foot by complimenting them on it and asking a question or two.

In summary, your preparation will probably determine whether or not you will win or lose in the negotiating arena.

17.6 EMOTION AND LOGIC

There are many similarities between the art of negotiation and the art of selling. In fact, a good negotiator uses many of the same skills and techniques used by good salesmen. However, there is at least one major difference.

For example, many salesmen will tell you that the buying decision is based on emotion and not on logic. If you recall, the purchase of a "super gas eater 8" over the "thrifty 4" was probably based on emotion—a feeling of importance or self-image as the purchaser sat behind the wheel. However, in my experience, negotiations are usually based on logic.

In essence, you sell your broad concept and probably make the sale based on emotion. For example, in industrial practice, your proposed partner may anticipate a promotion based on the success of your proposed venture. Then, logic is employed to a greater extent in negotiating the contract. After all, your proposed partner has to account to his superior for the overall deal, which includes all the terms of the contract.

Some technically trained negotiators overlook the importance of incorporating sales techniques during a negotiation. They base all of their discussion on logic. Don't make this mistake. All you need to do is to repeat your sales pitch from time to time during the negotiation. Many negotiations take months and, because of this, you may be competing with some new alternative. Besides, the longer the delay between the initial sale and a signed contract, the less likely it is that you will conclude a deal.

Consider the case of a young couple buying their first house in the northern part of the United States. They have selected the lot and a builder and agreed on the home. All they have to do is agree on the options and finalize the contract. Their problem is that they would like a fireplace and air conditioning, but can't afford both. They are emotionally swayed by the warmth of a fireplace, but are fully aware of its inefficiency as a method of heating.

The builder could point out that a fireplace would enhance the resale value of the house more than an air conditioner. In fact, in some areas, it is difficult to sell a home without a fireplace. The builder could add that, in this area, it invariably cools off at night and that it would be far easier to add air conditioning in the future, since he will prepare the house for air conditioning at no extra cost.

The builder is thus incorporating logic to sell a fireplace. A salesman would claim that the couple purchased it on the basis of their emotion, whereas a negotiator might argue that logic closed the deal.

In my own experience, logic is far more important in industrial negotiations. For example, assume that two companies, one Japanese and one American, are interested in a joint venture to develop, manufacture, and sell widgets. Each side wants to build the initial plant in their country. Both agree that they want to maximize profits.

The Japanese company points out the cost advantage of local production over importing, cost of shipping and duty, and their ability to close their borders once they have a local manufacturing facility. The American company would argue that the larger U.S. market justified a U.S. plant. A comparison of the potential profits leads to the most intelligent decision.

In one case, my client had developed a new process for separating hafnium from zirconium. He looked at the cost of the purified materials and felt that his company was entitled to a royalty that would give them 26 to 27 cents per pound. During the initial discussion, the prospective partner said that the cost of the high purity material was only 12 cents higher than the lower purity material, and that our process would use the lower purity material as feed stock. Obviously, we had to reevaluate our process and accept a royalty that reflected a percentage of the savings over the conventional process.

In general, you will not find a hard and fast rule for royalty negotiations. However, as a rule of thumb, a patentee with a strong patent should get 20 to 25% of the savings attributed to a patented project. On the other hand, arguing that a superior distribution system will result in greater sales is often effective in negotiating a declining royalty rate, such as 5% on the first million dollar of sales, 4% on the second million, and 2½% on annual sales in excess of 2 million dollars.

Actually, you will use both logic and emotion in negotiations. As you develop more skills, you will probably fail to distinguish between the two. However, don't forget to keep selling even after you have reached an agreement. This means that when you have a cooperative venture, keep your partner interested in continuing the project. In other words, when you have convinced management to fund the second phase of your project, keep selling, so that they do not change their minds.

17.7   TIMING

Don't overlook the importance of timing in negotiations. Timing includes the time of day, the length of the meeting, and the number of months to complete a deal.

To begin, if you are scheduling a meeting, select a time of day when you will be at your best. If you are a morning person, schedule the meeting for early in the day. On the other hand, if you can't get going before noon, pick a time later in the day. This is a small point, but it can be used to your advantage.

In general, you can plan for about 3 hours or less for an intercompany meeting. It seems that after 3 hours, the parties tend to go over and over the same ground. Therefore, you will do better to adjourn until the next day, review your position with the members of your team, and then reconvene the next morning. At times, you may want to adjourn for several weeks to develop additional data.

Licensintorg, a Soviet trading company, frequently schedules a week of negotiations and meets with the other party for a period of 3 to 4 hours each day. The system works well and is very efficient for detailed contractual negotiations.

At other times, a marathon negotiation may be in order. For instance, one session in Geneva, Switzerland with the president of a French company lasted from 9.00 A.M. to 9:30 P.M. At 9:30, the parties signed an agreement that ended multinational litigation. We were convinced that if we had taken a break, we would not have settled the litigation. In that case, the parties had agreed in principle at noon, yet could not agree on the details for the next 9 hours. At the end of the negotiation, we were also convinced that neither party would have conceded one additional dollar.

Many negotiators are optimistic about the timing for intercompany negotiations. For example, I recall a vice president of a major company insisting in September that a joint venture would be finalized by the year's end. Both parties wanted to go forward together, were completely cooperative, and reached a final agreement in early June.

As a general rule, patent licenses and joint ventures take about 10 to 13 months from the initial contact to a signed agreement. International agreements can take up to a couple of years. Thus, you should not be impatient or overly optimistic about concluding a quick deal.

17.8 SOME NEGOTIATING TECHNIQUES

Peter Casella is a good friend and an effective negotiator. I recall one session in which he saved the day. Peter had arranged for the new commissioner of patents to come to Buffalo to speak at a dinner meeting of the Niagara Frontier Patent Association. Our membership consisted of only 33 members, yet we worked hard and obtained 120 dinner reservations. The press was invited and a TV interview was scheduled.

On the morning of the meeting, the commissioner's assistant called to say that an urgent matter had come up and that it was impossible for the commissioner to attend. Peter set up a conference call to include me, and the commissioner's assistant repeated that it was just impossible. Peter replied, "I don't hear you Isaac." He then emphasized the TV interview, the number of attendees, the press and other arrangements, and convinced the commissioner to come on a slightly later flight, but in time for our meeting.

I still chuckle when I recall the incident and realize how important persistance can be as a negotiating technique. Peter just wouldn't accept a "no."

Good negotiators are seldom, if ever, in a hurry. In other words, good negotiators have patience. In some cases, negotiators have delayed for months, until they get an important concession. Usually, your only effective counter is an offer by another party.

Another common, minor technique, which is often overlooked, involves the seating arrangement. You may remember the lengthy impasse over the seating arrangements for discussions about the possible solution to the Viet Nam war. It may seem silly to consider seating as important, compared to the realities of war.

However, have you ever noticed the psychological effect when someone sits behind a desk and you are on the other side? I recall the first time I met Ray Tritten, who was president of AMF Incorporated. Ray got up from his desk and sat next to me. This simple act of courtesy told me a lot about Ray. He was people oriented and was a joy to work with.

Whenever possible, sit next to the person with whom you are negotiating rather than across the table. When there are several members on each team, split up, instead of aligning yourselves on each side of the table.

When you do sit on their side, don't try to look at their notes. Remember, the goal of most negotiations is to work together for a common, mutually beneficial goal.

There is one technique that is useful in solving an impasse. The problem is that the technique requires care to avoid giving the appearance of a concession. In essence, you say, "We seem to have an impasse. Why don't we go on and see what other problems we may have."

From this point try to resolve as many minor points as possible and then ask if there are any suggestions for compromising the final issue. The use of reason and an honest attempt to satisfy the other party's needs will often produce a solution.

In summary, a successful negotiation will satisfy the requirements of both parties and result in a mutually beneficial arrangement. A successful negotiator is considerate of the needs of his opponent and by preparation, diplomacy, and sensitivity, negotiates favorable arrangements.

18

Engineering
Supervision

18.1 INTRODUCTION TO SUPERVISION

Your ability to become a good supervisor does not rest on your technical ability, but on your ability to deal with people," according to Dr. Joseph V. Fiore, director of applied engineering at AMF's Morehead Patterson Research Center. Joe has over 20 years of supervisory experience and is well-thought-of by his peers, subordinates and superiors. Joe is well thought of because he motivates his team of engineers and scientists.

In fact, Joe says, "The key to being a good technical supervisor is an ability to motivate others." It is probably easy for Joe to motivate others, since he is a highly motivated individual. Joe enjoys his work, likes people, and is enthusiastic about his job and his profession.

You can become an effective supervisor by following Joe Fiore's example and by applying many of the rules that you learned in the preceding chapters. However, once you become an engineering manager, you will do even more, because you are responsible for a team of technically trained individuals. Part of your job is to mold your team into an effective organization.

Before going further, consider the problem of making the transition from a nonsupervisory engineer to a supervisory position. Joe Fiore says, "The difficulty in the transitional phase depends on the individual." He adds, "Experience as an officer in the military or in supervising nonprofessionals facilitates the transition." However, the transition can be fairly easy for anyone who "is mature in dealing with people."

According to Joe, his first experience as a supervisor was at the age of 22 when he completed officer's candidate school. On his first day as an officer, he put on a new uniform and sat down at mess, instead of standing in line. The service started him off on the right foot with an indication of his new rank. Joe says, "The military taught me the one best lesson about supervision. When you are in the field and come up for chow, your men go first." In other words, "Take care of your people first and be aware of what they do for you." Besides, "Your group is only as good as what you put into it."

Joe emphasizes the importance of motivation and of developing enthusiasm. He says, "You can develop enthusiasm by having a real love for your work, whatever the assignment." Even if you are reluctant, always develop enthusiasm. It is important, according to Joe, "to work within the framework of the corporate rules, with the proviso that you must have flexibility in order to deal effectively with fellow professionals."

As a general rule, "Treat others as you would like others to treat you."

18.2   COMMUNICATIONS

Have you ever analyzed the difference between a good supervisor and a poor one? In many cases, it is a matter of communications. A good manager tells his subordinates what needs to be done and how their work fits into the overall project.

Joe Fiore advocates an open door policy and recognizes the importance of communications in supervisory positions. Actually, there are three aspects to communications: within your group, with other supervisors, and with your superiors. Joe Fiore emphasizes the importance of communicating with the members of his team.

Joe says that one thing fosters good communication, "You must develop a talent for trusting those who work for you." For example, "If a subordinate develops data, don't question the raw data, question the conclusions." Whatever else you do, Joe warns, "Never cut off the lines of communication with one of your subordinates. If you try to discredit the data, you will probably destroy the lines of communications."

Another aspect of supervision, according to Joe, is to "keep up with the literature, since the engineers who are working at the bench may not have time to do so. Part of your job is to keep them informed about the state of the art."

Joe adds, "Never talk down to a member of your group, make them feel equal as a person." It is also important "to give each member of your team a pat on the back and recognition for their contributions." Joe also refers to the proper use of pronouns in effective management. In other words, you take the blame for any of the group's shortcomings and they get the credit for any successes. Joe says, "When someone asks, 'How come,' you answer with 'I,' but for credits, answer 'We.' "

Another mark of a good boss is "one who is always able to remind you that he is the boss without telling you so," acording to Joe Fiore. In effect, you ought to be "one of the guys and at the same time you are still the boss." Joe says that you can do this by "developing their respect for you, by becoming someone they can look to for guidance."

Joe goes on to say, "When you give a difficult assignment, take the time to work with them." He refers to this as "a complimentary type of thing that is the way supervision should work." In fact, Joe describes this concept with a family analogy. He says that it is a lot like being a parent, "You are part of the family and yet receive respect partly for age but primarily for guidance."

Bruno Miccioli was an exceptional supervisor, because he was people oriented and really cared about individuals. Bruno was the team quarterback; he was part of the team and carried far more than his fair share of any burden. Bruno was a receptive listener and an excellent counselor. He developed a partnership concept and included his subordinates in executive presentations. He got them exposure and gave credit where credit was due.

Bruno also met Joe Fiore's criteria for an exceptional executive. In Joe's words, "One yardstick for effective supervision is to count how many subordinates who have left come back voluntarily to visit because they like the atmosphere." Many of us who worked with Bruno kept in touch with him long after we left the company. All of us felt a terrible loss when Bruno died.

In general, a realtor will tell you that there are three key factors in evaluating real estate. They are location, location, and location. In supervision, there are also three key factors—communication, communication, and communication. When you communicate effectively, you manage effectively.

18.3 POSITIVE AND NEGATIVE FEEDBACK

"One of the best techniques for motivating individuals is a pat on the back," according to Joe Fiore. At the same time, you should "have an interest in each member of your team, his work and his output. Whatever else you do, don't ever squelch enthusiasm or put down an individual." Your job is to motivate the members of your team and to generate enthusiasm within your group.

Nevertheless, there are times when constructive criticism is needed. If you fail to provide constructive criticism, you will probably fail as a manager. However, try to avoid unpleasantness by being diplomatic. Your aim is to guide, not belittle.

One approach, which has been recommended by several personnel experts, is the sandwich technique. In other words, you praise an individual for something he did well, suggest the need for improvement in another area, and then praise him for something else that was done well. The sandwich technique is not always applicable; however, as a general rule, you should use praise at least twice as much as you use criticism to motivate people.

Timing is an important aspect of positive or negative feedback. When a child touches a hot stove, he feels pain immediately and learns that he shouldn't touch the stove. On the other hand, if he didn't feel pain for a week, he would have forgotten what caused it. Pain is a poor analogy, but does illlustrate the importance of timeliness in applying constructive criticism.

Some supervisors defer constructive criticism until the annual performance appraisal. Then, they overlook the problems and try to skip lightly over the evaluation with generalities. Don't make this mistake.

In reality, timely constructive criticism equals guidance. Think about what you might do if you were a quarterback who had been sacked five times by the same left guard. You wouldn't wait until the following week to suggest that your right guard take corrective action. Instead, you might suggest that he team up with your right tackle to give you a little better protection.

On the other hand, an annual performance appraisal can be an effective tool for supervision. In fact, a new supervisor would be wise to give quarterly or semiannual reviews. Emphasizing "where we are going and what we can do better" is an effective form of direction.

One serious problem encountered by many supervisors relates to an individual's less than satisfactory performance. Sometimes, that same individual has performed well for another supervisor. If this is so, the root of the problem may be a lack of direction. At times, all you have to do is tell an unsatisfactory employee what you want. In other words, you can help the individual to establish management by objective goals. On the other hand, there may be times when you fail to motivate an individual in spite of your best efforts. Once again, it is not necessary to have an unpleasant confrontation.

Assume that John persists in working on his favorite project and ignores a major project which is far more important to the company. You might say, "John, we have talked about the importance of the beta project several times, but still have a problem with your priorities. Apparently, I've failed to convey the seriousness of the situation. Therefore, in order to correct the problem, I am relieving you of all other responsibilities and want you to complete the circuit design for the beta project this month. If you have any problems with the design or encounter any delays, let me know and I'll help. John, we just have to complete the design work within the next 30 days. Let's review your progress and any problems again on Friday morning. I really think that we can get the job done on time. Do you agree or do you foresee any problems?"

A subsequent discussion might proceed as follows. "John, we are disappointed in the further delay on the beta project and believe that there is only one solution. We have someone else who will finish the job. It is obvious that you have a lot of talent, however, our solution requires that you look for another job. It's possible that you will find something within the company. For example, there is an opening at the Barlow Division and I've arranged for you to meet with personnel this afternoon to discuss the possibility of a transfer. Even if the Barlow position doesn't work out, you will have 3 months pay to tide you over until you find another job."

Terminating an employee is seldom, if ever, pleasant. However, if you have been effective in your role as a supervisor and consulted with a personnel professional, you can do the job diplomatically and minimize any unpleasantness.

When you terminate an employee, you should be prepared for one more difficult step. It is usually advisable to make the termination effective within a few days. In other words, separate the terminated employee from your group. The last thing you need is a malcontent within your team. Yet many employees want to stay until they find a new job. The solution is to say something like, "You will be on our payroll for 3 months, and personnel will arrange to pick up your calls and help you in making the transition."

In my experience, it is relatively difficult, if not impossible, to force an individual to change. Therefore, when you encounter one who is a serious problem, talk to personnel. Of course, first you should discuss the matter with the individual and make certain that he understands what you want. When all else fails, suggest that the individual accept a transfer or find another job.

18.4 BUILDING A TEAM

A team that works together will accomplish far more than a group of individuals. According to Joe Fiore, "The only way to develop a team is when each member of your team has his own role." For example, a good quarterback does not try to do the fullback's or lineman's job. He recognizes that the fullback is probably better equipped to pick up a few yards through the line.

Joe Fiore says, "Try to complement what you have with the abilities of your subordinates." Joe adds, "It is critical, you must accept talent in areas where you are not sufficient." Hence, "You should not try to outshine your subordinates in all of these roles. If you do, you don't need subordinates, because you are doing their jobs."

When you build your team, Joe advises, "don't inhibit creativity." He says, "You should avoid oversupervision, because if you make all of the decisions or think that you do better than your subordinates, they'll stop making decisions." What you want to do is "treat them in a way that allows them to be individualistic, as long as it conforms to the overall program's objectives." In other words, provide an atmosphere that "fosters freedom of thought."

Jean Hackenheimer, a former manager of legal administration at the Carborundum Co. leads by example. She works harder and longer than any of her subordinates. She is also strongly opposed to inefficiency. No one ever needed to ask one of Jean's group to pitch in, because she let everyone know about the department's goals and priorities. She established a team effort to meet those goals.

18.5 COMPETITION AND PARTICIPATIVE MANAGMENT

One corporate president established two executive vice presidents and set them up to compete against each other for his job. There was one serious problem. The jobs were mutually dependent and the competitive nature of the two individuals did not lead to a harmonious relationship. In fact, their lack of cooperation was detrimental to the company.

Competition may be beneficial in a few instances, but more often than not leads to dissatisfaction and waste. After all, everyone should be working together for the good of the company rather than concentrating on some psuedocompetition.

Nevertheless, some organizations are based on developing internal competition. For example, many of the very large law firms actively promote competition among the associates. They hire 8 to 10 associates with an expectation that only 1 will become a partner. The concept is that only the best and hardest working will become a partner after 8 or 9 years. The result is that most of the associates work incredibly long hours, become very good lawyers, and leave the firm.

The concept of participative management has not been well-accepted, but is frequently proposed by industrial psychologists. In general, the concept seems contrary to most management techniques. For example, how much respect would the crew of a sinking ship have for a captain who asked for a vote on whether to abandon ship?

However, you can use participative management to a degree for the professional development of your staff. To begin, you should know the technical strengths of your subordinates. Those with the technical strength for an advanced degree should be encouraged to pursue further study. Recognize and cultivate those with the potential for management.

One way to develop a potential manager is to delegate increasing amounts of responsibility. Let them make decisions and do your job during your absence. Actually, this isn't really what is meant by participative management, but it is an effective way to motivate and develop individuals with management potential.

Joe Fiore tells a story about a subordinate who wanted to become a group leader. Joe recommended that he be given a chance, but it was decided that the individual did not have what it takes. A short time later, the subordinate left the company and went on to become a successful executive. "At times, you just have to take a chance and give someone the opportunity to grow," according to Joe Fiore. After all, "another mark of a good supervisor is an ability to develop his own replacement."

In summary, supervision is a complex subject and is even more so in a multifaceted group. It's up to you to be flexible, to work with the members of·your team, and to help them get the most out of their talents. As Joe Fiore suggests, "You can't put yourself on a pedestal. In fact, you can avoid a lot of problems if you will leave your desk and go out into the lab, go out to the places where your subordinates are working. Treat others as you would like to be treated, and you will succeed."

18.6 MAKING EFFECTIVE FORMAL PRESENTATIONS

In your role as a technical manager, you will probably be called on to make formal presentations to higher management.

One of the best courses that I've taken taught me a lot about making effective presentations. The course was a writing seminar given by Barry Tarshis, author of *How To Write Like A Pro*. Barry taught me a lot, but one lesson overshadowed all of the others. Barry repeated over and over again, "Take a reader oriented approach." That rule is equally applicable to oral presentations. All you have to do is substitute listener for reader. Actually, most of what Barry taught me can help you to make better presentations (see Section 3.2 in Chapter 3.)

Bruno Miccioli was a master at making effective presentations, because he took sufficient time to carefully prepare and stage each presentation. Bruno also involved his subordinates and used visuals and rehearsals to develop a polished presentation.

Joe Fiore also emphasized the importance of involving your subordinates in developing a formal presentation. He points out, "You should encourage your subordinates to attend professional meetings, to write technical papers and present their findings to a suitable forum." He says, "You should be aware of their capability to communicate and help them to overcome any deficiencies in writing or presenting data. However, be cautious, don't try to force someone who is reluctant."

Barry Tarshis and Joe Fiore both had one thing in common with Bruno Miccioli. They are dynamic speakers, because they use a number of techniques that help them to make more effective presentations. Fortunately, many of these techniques are easy to learn and will help you to become a more effective manager.

Overcoming Nervousness

Almost everyone feels nervous before making a formal presentation. In fact, many accomplished speakers who are well-prepared have hands that shake so badly that they can barely hold a water glass. It may seem unbelieveable, but you can use this nervousness to your advantage. All you need to do is channel your nervousness into your presentation.

First, stand comfortably with your weight equally divided between both feet. Lean slightly forward, but keep both feet firmly planted. This comfortable stance will help you to avoid shifting your weight from one foot to the other, rocking on your heels, or some other nervous fidgeting that will distract from your presentation. Watch others when they make presentations and note how shifting or pacing distracts your attention from the subject matter. These visual distractions can reduce your effectiveness. Fortunately, they are relatively easy to correct.

Your second step is to move away from the lectern. I was a lectern clutcher for many years and was horrified by the thought of making a formal presentation without one. Besides, I asked myself, "What can I do to hide my shaking hands if I can't hold on to the sides of the lecturn?"

As an engineer or scientist, you know what would happen to an airplane wing that didn't flex. You wouldn't design a rigid wing, yet many of you will try to eliminate the slightly shaking hands. It is analogous to Jack, a young mechanical engineer who bought a used washing machine. It was an old Bendix front loading machine that vibrated so badly, it moved across the basement floor. Jack bought four large bolts and bolted the machine to the floor. The machine still vibrated and the basement floor developed four very large cracks. If you tie down your shaking hands, your nervousness will probably appear someplace else.

Actually, there is a simple solution for overcoming this type of nervousness. You will find that when you use your hands the nervous tremor disappears. The fact is that a nervous tremor is probably imperceptible to the audience, and when you move your hands to make gestures, any slight tremor disappears. Besides, gestures will make your presentation much more effective.

Many engineers use a blackboard, which is effective for informal presentations. However, you can apply this same principle to a flip chart, overhead projector or slides for formal presentations. Get involved with your presentation, walk to the screen, pass out samples and use your hands to illustrate high, low, large, small, and so on. Just be natural and overcome your tendency to fix your hands to a lectern.

A Positive Attitude

You can develop a positive attitude toward your presentation by organization and preparation. Decide what you want to accomplish, organize your thoughts, support your thoughts with facts, and be enthusiastic.

Don't overlook the importance of appearance in making a formal presentation. I recall one effective presentation in which an engineer did a creditable job. The only trouble was that he had an overstuffed wallet in his back pocket and every time he moved, you expected his pants to rip. It only takes a few seconds to place your wallet, keys, and change in your desk or brief case. Getting rid of them helps eliminate a tendency for fidgeting.

Whenever possible, videotape your rehearsal and ask for constructive criticism. In many cases, you will want to videotape a proposed presentation three or four times. Practice your gestures, practice making larger gestures, and build confidence as you polish your performance.

Visual Aids

You need visual aids for your presentation, even if it is a short talk. Simple visual aids are usually better than complicated ones, and using them will help you to make a better presentation. First, visuals are effective. Studies have shown that people remember a picture better than the spoken word, and a simple illustration can help you make a point. Second, good visuals build confidence, because you know that you have something to work with. They also exhibit preparation and tell your audience that your presentation is so important that you prepared visuals to illustrate the more important parts.

In addition, one reason for utilizing visuals overshadows all the others. Visuals get you involved in the presentation and thus help you overcome any nervousness. They compel you to move to the screen and make large gestures. In fact, good visuals will make you more effective.

You might ask, "What is a good visual?" A good visual can be a diagram, a simple graph, or at times a few key words. Flip charts, transparencies for an overhead projector, and slides can also be effective. They will have greater impact when you use color. Think back to the days of black and white television. Wouldn't you rather see a program in color?

The best visual differs somewhat from the ones already mentioned. It is a sample or prototype that allows you to involve your audience. What does a good car salesman do? He puts you behind the wheel. The effective computer salesman sets you down at the keyboard. When you involve your audience, you have their attention. When you have their attention, you are on the road to success.

In October 1969, I was program chairman for a meeting of the Licensing Executives Society in Daytona, Florida. Gerald Nierenberg, author of *The Art of Negotiation*, was a key speaker. Gerry, a dynamic speaker, used visuals and about midway through his presentation, he invited a large group up to the stage to participate in an experiment. I thought that it would be a disaster and that with so many people on the stage, he would lose control of the audience. He didn't, and 15 years later, I remember the standing ovation at the end of Gerry's presentation.

Techniques For More Effective Presentations

You can become a more dynamic and more persuasive speaker with a little practice. It is true that many of us do not have the rich, dynamic voice of Gerald Nierenberg or Barry Tarshis. Yet you can apply many of the same techniques that they use to become more effective speakers.

For example, the Rev. Robert McBride (deceased) was one of the most dynamic speakers that I have ever heard. He had a rich voice and was a master at using two techniques that are frequently overlooked by less experienced speakers. One is eye contact and the other is a pause.

Eye contact is probably the most important technique for making an effective presentation. What do you do when you talk to an individual? Do you look at the corner of the room? Of course you do not. You look at the individual and observe his reaction. You can do the same thing when you speak to a group. Just vary the eye contact by moving from one individual to another. Try to hold eye contact for about 5 seconds and then move to someone in a different part of the audience. When you look someone in the eye, you have their attention.

Father McBride would ask a rhetorical question, establish eye contact, and pause for up to 5 seconds. When he did this, you focused your attention and couldn't help but think, "The next question might be directed at me." He forced his audience to think about what he said.

Good speakers like Father McBride use pauses to get your attention. Actually, even inadvertent pauses can liven up your presentation. They offer a dynamic change of pace, focus attention, and help you develop confidence. After a question, pause and hold eye contact until you see a nod or a head shake. It may seem like hours, but is probably no longer than 1 to 2 seconds.

I recall the first time that I tried this technique during a course on making effective executive presentations. I looked at Jim Emmett and after what seemed like minutes, he nodded. Afterwards, I asked Jim, "What took you so long?" When I reviewed the television tape, we timed the pause and it had only lasted for 1½ seconds.

There are other techniques that will help you to become more persuasive when you make an executive presentation. For example, no one likes to listen to a monologue, yet many inexperienced speakers fail to remedy this problem. You can overcome it by changing the volume of your voice. At times, drop your voice and speak softly. Vary the speed of your voice by picking up the tempo (not too fast) to show enthusiasm and then pause to make a point.

Whatever else you do, eliminate nonwords like "ah." Have you ever sat through a presentation and counted "ah's?" I have heard some people use so many "ah's" that I lost track of all else. In the previously mentioned course, the audience was asked to snap their fingers everytime a speaker used a nonword. After a few corrections, the speakers learned to avoid nonwords.

You can also become more persuasive by eliminating qualifiers such as "I think" from your presentations. Recite facts, not guesses. When you say "I think," you are telling your audience that you don't know or aren't sure. You should even avoid "I believe," except in those situations when you are asked for an opinion.

When you videotape your rehearsal, try to eliminate any visual distractions. Look closely to see if you drop your eyes to your notes before you finish the sentence. If you do, try again but complete the sentence and then look down. The resulting pause will not be noticeable. Besides, the effect of speaking while looking down at your notes can be devastating.

I recall videotaping one presentation in which I used a number of props. I continued to speak as I walked to the table and picked up the sample. My instructor advised that I stop talking, walk over and pick up the sample, and then continue. The difference was very dramatic. Try to repeat this simple procedure when you videotape one of your presentations.

In summary, you can become a more effective speaker by being listener oriented and by eliminating visual distractions. Involve your audience and use eye contact, voice modulation and pause to make your points.

Questions and Answers

In an earlier section of this book, you learned about the one universal answer to difficult questions. However, "I don't know" isn't always appropriate for an executive presentation. It is true that you can offer to find out and report back, but don't overuse this approach.

In general, your best approach is preparation and anticipation of likely questions. However, as you encounter larger audiences, you may find it helpful to control the questions.

For example, when you address a large group and agree to answer questions, it is helpful to repeat the question. Repeating the question gives you additional time to think of an answer and avoids misunderstandings between the question asked and the one answered. At times, you may want to condense the question or provide a partial answer.

When you answer a question, establish eye contact then shift to another person, rather than allowing the person asking the question to take control.

As Joe Fiore advised, "Encourage the members of your team to write and present technical papers." You should also present technical papers and develop your own skills in making effective presentations.

19

Some Special Problems for Women in Engineering

After years of discrimination, women are finally beginning to advance their careers in many professions at a pace equal to that of men. It's true, of course, that prejudice against women in business still exists, but a combination of factors—the women's movement, affirmative action, and the increasing proportion of women graduating from college with professional degrees—is helping more women than ever before to reach higher levels in corporate management.

But what does this mean to women interested in an engineering career? First, women today face a tougher career path in engineering than in other professions, but this doesn't mean that there aren't opportunities. It's largely a matter of choosing the right company. More and more corporations are aggressively seeking bright women graduates with engineering degrees, and more and more corporations are providing management training and real opportunities for promotion for these women. Unfortunately, when it comes to engineering positions, some corporations are simply paying lip service to affirmative action programs and have little or no interest in promoting women into higher management. Questions about management development programs and the number of women in higher management will help to clarify their real interest.

Opportunities for management positions are there for the career minded woman engineer. But if she wants a career in engineering management, she has to work harder and smarter than her male colleagues. She must also win recognition for her accomplishments and apprise management of her career goals.

19.1   ESTABLISHING AN IDENTITY

You can establish an identity as an individual with high potential by following the suggestions given in earlier sections of this book. However, a woman in engineering will probably have to convince management that women should be taken seriously.

In 1973, I participated in a U.S. delegation to the Soviet Union headed by Dr. Betsy Anchor-Johnson (a physicist and an Assistant Secretary of Commerce). Betsy established a reputation for initiative, hard work, and an ability to make things happen. Because of her efforts, the delegation made a significant contribution toward improving relations between the two countries.

A few years later, I was again privileged to work with Betsy when she initiated an exchange with other Eastern European countries. By applying industrial practices, Betsy Anchor-Johnson overcame bureaucratic red tape and completed her program on schedule. In her work for the U.S. Government, Besty proved that a woman scientist can succeed in top management.

Today, Dr. Betsy Anchor-Johnson is a vice president of General Motors.

19.2   THE IMPORTANCE OF FAMILY SUPPORT

Dr. Edgar H. Schein, professor of management at the Sloan School of Management, M.I.T., says that "from about age 25 to 40, the family colludes with the primary career occupant to build a successful career." Professor Schein goes on to state that "the children are kept out of the career builder's hair while he or she is busy."*

It is true that our culture is based on a family unit. That family unit includes a primary wage earner and a partner who makes sacrifices to enhance the career of the primary wage earner. The problem for female engineers is that most engineering managers are male and assume that these women will revert to a traditional wifely role. Male managers frequently fail to perceive a change in society. Even though some male managers respect a woman engineer for her ability, they may fail to develop her managerial talents.

Your emphasis on career planning can help you to overcome this type of managerial myopia.

19.3   WORKING SMARTER IN CAREER MANAGEMENT

"Don't be bound by a job description," warns Julie Fenwick-Magrath, manager of human development at AMF's world headquarters.

*Reprinted from *Improving Face to Face Relationships* by Edgar H. Schein, Sloan Management Review, Winter 1981, Volume 22 Number 2, pp. 49–50 by permission of the publisher. Copyright 1981 by the Sloan Management Review Assoc. All rights reserved.

Julie, a bright articulate manager, suggests "gaining visability with senior management." She also suggests that a career oriented individual "makes the effort to obtain information directly from a top manager, rather than being satisfied with second hand information handed down by a superior."

After 8 years in management development, Ms. Fenwick-Magrath says that women have to work harder than men to succeed in business. She suggests that "women must be willing to make more sacrifices in their social life," and that it is important early in your career to have the mobility to take advantage of a career opportunity.

Changing jobs can contribute to your professional growth, however, Julie cautions a woman "to watch out that you are not being set up for failure." She also suggests that "a woman should ask a potential employer about career opportunities," and stresses the importance of lateral transfers to gain broader experience. You can improve your chances for success if you make a thorough study of your potential employer. Try to find out if other women have been given real responsibility and talk to them about their success. Discuss affirmative action with personnel and find out about the likelihood of lateral moves for career development.

Julie advises that "you ask potential employers how many women they have in management and at what level." She would also question "if they would accept a woman in a particular role." For example, are you being asked to serve as a staff consultant to line managers who will be unwilling to take your advice? When you do consider a job change, Julie suggests that you ask your potential boss for references and that you check the references before accepting a position.

Julie Fenwick-Magrath's attention to career management and hard work have helped her to establish an image as a corporate manager with exceptionally high potential.

19.4   OVERCOMING ISOLATION

Lena Hudson, a delightful person, retired as a supervising engineer. Lena was also a leading lens designer with a number of patents to her credit. I wrote a number of her patents, as well as a number of patents for the men she supervised.

In my dealings with Lena Hudson, I recognized that she was thorough, very bright, an expert in her field, and a good supervisor. Nevertheless, she had two problems. First, she was an engineering supervisor in the early 1960's, when it was extremely rare for a woman engineer to be given a supervisory position. Lena's second problem was that she had become a specialist in lens design. In doing so, she had not gotten the more diversified experience of her male colleagues and, therefore, was not considered for a higher position.

Lena is a fine person, who made a significant contribution to her profession. She also helped to train a few of the bright young men who learned much from her talent and experience. In today's environment, Lena Hudson would have greater opportunity to achieve a higher management position. Nevertheless, it would be her responsibility to gain diversified experience and to prepare herself for further promotions.

Jean Hackenheimer, an example of a remarkable manager, was voted the outstanding woman in management on the Niagara Frontier in 1981.

Jean's entire career was spent in patent and legal administration. She served as a legal administrator for a number of years and continued to assume broader and broader responsibility. Jean used her organizational ability to produce one of the most efficient and effective units in the corporation.

Jean always worked hard, put in extra hours, and demanded a full day's work from her subordinates. She is also one of the most persevering individuals I have ever known. For example, after the personnel department refused to suggest information on management development, Jean attended several university courses for women in management, joined professional organizations for women managers, and initiated informal luncheons for women managers in her own company.

After years of specialization, she became an exceptional manager and supervised the financial, personnel, and administrative operations for a legal division. Jean is an example of a woman who, on her own initiative and through hard work, was able to break out of her specialization and broaden her background. It was the key to her success. If you follow her example, you will be more successful.

19.5 OVERCOMING THE OLD BOY SYNDROME

The Niagara Club in Niagara Falls, New York, like many other private clubs, excludes women from the main dining room during lunch. It typifies the male dominated clubs that keep women out of the "inner circle" and give male competitors an unfair advantage.

Nevertheless, you can overcome this type of impediment, if you follow the example set by Joan Norak. Joan is a fine lawyer who practices in Chicago, Illinois. I worked with her on two lawsuits and was impressed by her legal proficiency, technical competence, and ability to communicate with corporate managers. Joan didn't break the tradition at the Niagara Club, but did participate in business meetings held in other establishments. By becoming an integral part of a team, she at least partially overcame the "Old Boy Syndrome." Joan served her apprenticeship with a leading trial lawyer, and like all young associates, transported massive legal files and worked long hours without complaint. She also developed a reputation for doing high volume and high quality work. Joan did not become a member of one of the male dominated luncheon clubs; however, she was accepted as an equal in a predominantly male dominated profession.

19.6   FACE TO FACE COMMUNICATION

A sense of your own identity, and an ability to decipher other people's values, are key elements for effective face to face communication, according to Edgar H. Schein. Professor Schein states that "negotiations require great sensitivity, humility, self-insight, motivation to solve the problem and behavorial flexability."*

Many women are trained in skills of listening from an early age; their male counterparts are trained in the importance of dominating others through sports. The result is that women are far more sensitive than men in detecting key points during a negotiation.

As a woman, you can become a better negotiator than your male colleagues and, in fact, better than many of the top executives. Don't forget the observation of a leading trial lawyer that, as a general rule, "the higher the manager, the poorer the negotiator." One reason why many top executives are not better negotiators is that humility and behavorial

*Reprinted from *Improving Face to Face Relationships* by Edgar H. Schein, Sloan Management Review, Winter 1981, Volume 22, Number 2, p. 47 by permission of the publisher. Copyright 1981 by the Sloan Management Review Assoc. All rights reserved.

flexability are contrary to the male's macho image and are traits not usually associated with top management. If you develop these traits, you can surpass many, if not most, of your competitors.

The intelligent use of diplomacy and aggressiveness can help you to succeed. When you couple aggressiveness and diplomacy with effective communications, you enhance your image and your opportunity for promotion.

19.7 SUMMARY

The women mentioned in this chapter succeeded because they established their identities. All are enthusiastic, work hard, and receive recognition for their contributions. They succeeded because of their planning, the effective use of their time, and their career management. Above all, they sold themselves and their ability to meet their goals. On the other hand, if you are not promoted, or fall behind in salary advances, ask yourself if you have faithfully applied the principles set forth in this book. Ask your boss for a candid appraisal of your performance and an assessment of your career potential with your present company.

If you are not satisfied with your appraisal or your opportunity for promotion, consider your alternatives. If your present employer is not fully utilizing your talents, read the next chapter and find another job. If, on the other hand, you are convinced that management has repeatedly promoted men of less ability than yourself, and are willing to risk future advancement, consult a lawyer. You may have the basis for a sex discrimination lawsuit.

Before filing a suit, consider your alternatives carefully. Recognize that suing your employer may be harmful to your career, but if you feel it's necessary, do it. Even without suing, the threat of litigation may help you to achieve a higher position. Again, don't overlook the danger in this approach. Discuss your situation with a lawyer and use diplomacy. In general, you should try to negotiate an equitable solution.

Bear in mind that a threat of litigation is tantamount to a threat to leave the company. Threats should be used rarely, if at all.

20

Changing Jobs

At times, drastic action may be required to advance your career. For example, if you were passed over for promotion more than once, or if your company was acquired and promotions were preempted, you may want to find a new job. The sad point is that many successful engineers and scientists don't really know how to find a better job. It is sad, because it can be relatively easy to change jobs and at the same time to obtain a promotion. As a first step, consider the points raised in this chapter.

Before you look for a new job, review your career goals and your experience. Analyze your annual performance appraisals to determine your strengths and weaknesses. Ask yourself, if you can benefit by staying with your present company while you complete an advanced degree? Completion of your major project might allow you to publish an important paper or list a major accomplishment on your resume.

Robert Townsend, author of *Up the Organization*, suggests that you should not stay in the same job for more than 5 years. This may be true, but should be tempered if you are given broader responsibilities and training that will help you to reach your career objectives.

Besides, finding the right job is difficult and takes time. The alternative is to accept a less desirable position that will not advance your career.

Don't change companies for a few thousand dollars a year unless you are promoted. Don't repeat your previous experience and end up with 1 year's experience several times.

The size of a company should be evaluated in considering your needs. A larger company may have greater resources and more sophisticated equipment and offer an opportunity to work with specialists in many fields. On the other hand, a smaller company may offer greater responsibilities and an opportunity to work with top management. The company size should be consistent with your personal preferences, because you will be more effective in a comfortable environment.

Take your time in finding the right job and you will advance your career.

20.1   TIMING AND PREPARATION

As a general rule, never quit your job until you have another one. Some years ago, I agreed to accept a position with a small company and shook hands with the corporate vice president. I asked that he confirm our agreement in writing and refrained from giving notice until receiving his letter. The letter never came.

Carl Boll, author of the excellent book *Executive Jobs Unlimited*, suggests that you devote full time to looking for a new job. Based on my own experience, it may take 12 to 18 months for you to find the right position. This period of unemployment could have a devastating effect on your savings, pressure you into accepting an unsuitable position, and create a hiatus for a future resume which might cost you an important position.

Be prepared for a long search, take your time, and find the right job.

20.2 YOUR RESUME

"You should avoid functional resumes," according to James Emmett, manager of personnel at Clairol Inc. In general, a company is interested in your accomplishments, not in a list of responsibilities. "I want to know what you have done" says Emmett. However, he cautions against "trying to tell someone everything you have ever done." Jim also suggests, "When you answer an advertisement that requests salary information, include your present salary or a range you consider satisfactory."

An example of a resume that illustrates accomplishments follows:

As a senior engineer at Weston Electric Company, I:

- Initiated a cost reduction program that saved $700,000 per year;
- Installed a fully automated machine assembly plant utilizing 10 robots;
- Increased production in radio assembly by 300%; and
- Designed a microwswitch that produced annual sales in excess of $5 million.

During my 5 years before Weston, I was an associate engineer at Circle Electric Company. At Circle, I:

- Directed the quality assurance program for a circuit board manufacturing operation with $10 million annual sales;
- Supervised the quality control lab, which employed 7 technicians; and

- Sold a $3 million development contract to the U.S. Department of Defense.

Don't forget to direct your resume to the reader. It sounds simple, yet many resumes include a hodgepodge that is only meaningful to the writer. It is usually better to put your most recent accomplishments first, since these are of primary interest to the reader.

Remember, if you don't get the reader's attention in the first paragraph, he might not read any further. After all, the reader may have to scan several hundred resumes for one job. You can make your resume stand out by listing significant accomplishments in clear and concise terms.

Avoid clutter by using short action verbs, as suggested by William Zinser in his book *On Writing Well.* You should avoid using adjectives such as exceptional, outstanding, and superior since they do not provide meaningful information to the reader.

Whenever possible, your resume should be tailored for a specific position and for a specific company. For example, it is probably better to send a letter in response to an advertisement, rather than relying on a printed resume. Nevertheless, your resume would form the basis of the letter after an opening paragraph. The following example shows how easy it is to put your resume into better form.

This is in reply to your advertisment for a supervisory engineer with a background in electronics.

As a senior enginner at Weston Electric, I:

- Initiated a cost reduction program;
- Installed a fully automated . . . , etc.

Your letter (resume) can then be modified to emphasize those characteristics mentioned in the advertisement.

A resume should be brief and limited to no more than 1½ pages. Don't use small type or narrow margins. Remember that a resume is analagous to an advertisement and is intended to get you a job interview. Note also that if you include too much detail in your resume, you will not get an interview.

If you use a letter approach as opposed to a printed resume you will lose space at the top of the first page, so that your one page resume will grow to about 1⅓ pages. Avoid the temptation to fill the second page and remember that a *concise*, hard-hitting resume is best.

Two sample resumes are included in the appendixes to suggest different approaches for presenting your credentials. The resumes are probably a bit too long, but both emphasize accomplishments as well as pertinent educational and personal information.

20.3   FINDING A JOB

There are many ways to find a new job. Some may seem impractical, yet in given circumstances, have proven to be effective.

"More new jobs are found by networking than by all other approaches combined," according to an estimate by Jim Emmett. Jim defines networking as using your contacts with friends and associates. He admits that this approach is probably far more effective for an engineer with 10 years experience than for a younger engineer. As a young engineer, you can prepare for the future by keeping a 3 × 5 card file on all of your contacts. Keep updating your file and try to keep in contact with all of your associates.

Jim believes that executive recruiters, direct mail campaigns, and answering ads probably follow in that order as effective ways for finding a new job. However, Jim suspects that you will find a job more quickly if you write directly to companies.

"If a recruiter calls you," Jim advises that "you take 45 minutes to an hour for a face to face meeting even if you are not ready to change jobs." He also suggests that "you develop four to six contacts with executive recruiters who are active in placing engineers in your field or geographical area." These contacts can be very helpful for the future.

An engineering supervisor with an opportunity to hire engineers should use a recruiter, whenever possible, according to Emmett. If you are in this position, "Take advantage of the situation and interview several recruiters." Besides, you can spread the work around and develop a good working relationship with a number of professional recruiters in your area. When you need help finding a future job, they'll remember.

Women may have an advantage in relocation, according to Jim Emmett, if they utilize organizations such as Catalyst in New York, Inc., a nonprofit group that maintains an office at 14 East 60th Street, New York, New York 10022. There are similar groups in other cities that provide job and career information.

You should review help wanted sections in the *Wall Street Journal*, the *New York Times*, and professional magazines as a first step. Also, buy a copy of the *Wall Street Journal National Employment Weekly*. A problem is that many other engineers are doing the same thing and as many as 350 or more engineers may apply for the same job.

However, you can tailor your letter to meet the needs of a specific advertisement and improve the chance for an interview. Even if you are not interviewed, review your letter, polish it, and don't become discouraged. In one case, the personnel manager of a major glass company eliminated all resumes that did not include experience in the glass industry and an M.B.A., even though neither requirement was mentioned in the advertisement.

Nevertheless, you can find the right opportunity and then sell yourself. Contact engineers who have changed jobs recently, as well as business associates, sales representatives, former classmates, and friends. One recently relocated executive was surprised by the help he received from associates he had not seen in many years.

Edward C. is a design engineer at a major corporation. He had been working in a city where corporate personnel departments had an agreement against hiring from other companies in the same city.

Ed wanted to change jobs, but didn't want to move to another city. He asked an engineer at another company to set up a meeting with the engineer's boss. Ed had two Saturday morning interviews and accepted a new position before any contact was made with the personnel department.

Carl R. Boll suggests sending hundreds of letters written in the first person to potential employers and adds that these letters should be directed to an appropriate officer, such as the vice president of engineering, for best results.

Many job seekers object to the direct main campaign and pursue some other approach for finding a new job. At age 27, I placed 1 advertisement in a professional journal, received 29 replies, had 5 interviews, received 2 offers, and accepted a new job. Ten years later, I placed a similar advertisement in the same journal and received 2 replies and 1 interview.

ʹAn executive recruiter is a professional in matching people and jobs. In reality he is like a real estate broker. Many recruiters work on a contingency basis and don't get paid unless they fill a position. However, if you have good credentials, many recruiters will try to place you with a client. In such cases, an executive recruiter will help you to sell yourself and can give you valuable information about companies and their executives.

In my own experience, executive recruiters have been very helpful in changing jobs. The association of Executive Search Consultants, Inc., 30 Rockefeller Plaza, New York, New York 10012, will provide a list of their 56 members for a fee of $3.00. You can also obtain a list of over 2000 executive recruiters by writing to Consultants News, Templeton Road, Fitzwilliam, New Hampshire 03447. Don't be intimidated by the number since many of the recuiters specialize in sales executives, financial executives, and specialists in data processing. Nevertheless, if you send out over 1000 resumes, you will get results.

20.4 CONDUCTING AN INTERVIEW

A good interview is the result of careful preparation. Many years ago, I answered a blind advertisement, was called for an interview, and was asked to be there in 2 days. The interview proved to be a disaster. The executive knew very little about my field and didn't know what he wanted or how to conduct a meaningful interview. My lack of preparation contributed to a mutual understanding that this was not the right job for me.

Before going to an interview, obtain a copy of the company's annual and quarterly reports and its latest 10K (filed with the SEC). Call your stockbroker and ask him for information about the company. Check your library for recent articles on the business and review all of the information carefully.

At the library, select several books on effective interviews and look for a list of typical questions asked in preemployment interviews (such a list is included in Appendix L). Write out your answers and review the questions and answers several times. Review them once more immediately before the interview. Write out all of the questions you have about the present opening, the company, and the individuals with whom you plan to talk. A list of basic questions is included in Appendix M. Prior to the interview, try to obtain the names and positions of each individual you will meet. Check the library to learn if they have published articles or are included in *Who's Who In American Business.* If you find an article, compliment the author and ask if he has pursued these studies and also discuss his experiments. You will probably receive an offer.

Your questions about management development programs should be directed to personnel. You should also ask personnel to give you an overview of company policies and benefits. Most personnel departments will give you brochures that can be studied at your leisure.

You should prepare yourself for an interview with a trained psychologist or a personnel specialist. A typical interview by a psychologist or specialist will take 1 to 2 hours and is referred to as an in-depth interview. These interviews are designed to find individuals who are compatible with the company and its managers.

There are two types of in-depth interviews, according to Jim Emmett, "the more common one is designed to probe your background and to determine that you are competent in your field of expertise." The second type, a psychological interview, may be conducted by some companies to evaluate motivation, drive, self-discipline, and a number of other psychological factors that might affect your performance.

The basic rule in this type of interview, and in all others, is to be candid. Be prepared to answer questions about your strengths and weaknesses in a manner that emphasizes your strengths and minimizes weaknesses. A review of the first two chapters of this book will prepare you for elaborating on your strengths. Weaknesses can be tempered by statements like, "At times I may be a bit impatient or not overly tolerant of individuals who fail to do their job."

Jim Emmett suggests that you "ask questions during an interview," and says that "companies like people who probe a lot." Avoid asking dumb questions such as, "How much vacation will I get?" or "Do I qualify for a reserved parking space?" If you are concerned about such things, inquire about adequate parking or whether the plant shuts down for a scheduled vacation period.

In general, be enthusiastic about work, but don't appear to be a workaholic. Be well-prepared and candid. Inquire about opportunities for management development and if there is a policy to promote from within. After all, if the company is not interested in your management potential, you should look elsewhere.

20.5 SUMMARY

In finding another job, you should concentrate on getting a promotion as an inducement to move. You may find that getting a promotion at the time of changing jobs is very similar to getting a promotion at your present company. You will get a promotion if you have patience and work hard.

Don't· forget that "image is important," according to Jim Emmett. Wear your best suit instead of a sports jacket, have your shoes shined and don't wear white socks and safety shoes to a job interview.

In reality, finding a new job is very much like working for a promotion. If you do more things right than your competitors, prepare yourself carefully and apply the suggestions in all of the preceding chapters, you will be successful.

21

Conclusion

As an engineer or scientist, you do have ability and an opportunity for advancement, if you are willing to apply yourself and master the suggestions in the preceding chapters. In fact, you really have an advantage over other employees who do not have technical training. On the other hand, if you sit back and watch others be promoted, you have only yourself to blame.

If you want to be the vice president of research or engineering, or even president of your company, apply yourself in your present job and work hard to get promoted.

Remember, your future depends on your accomplishments and on obtaining recognition for those accomplishments; your future depends on career planning, goals, direction, and hard work; your future depends on your use of your company's resources to advance your career; and your future depends on selling your ideas and yourself.

For those of you who are willing to work hard to make it happen, there is one final step in achieving success. That step involves a small sign which is placed in the top drawer of your desk. The sign should be visible whenever you open the drawer. My sign is abut 3 × 7 inches, printed in india ink, and placed in the right-hand corner so that it never becomes obstructed.

Each morning when you arrive at work, unlock your desk and look in the top drawer. Each afternoon when you return from lunch, peek into the top drawer. At any time that your mind wanders from your work, open your top drawer. The sign's advice is difficult to follow, but will guarantee your success. It reads, "I must do the most productive thing possible at every given moment."*

Remember, it is up to you to make it happen.

Vaya con Dios.

*Hopkins, Tom, *How to Master the Art of Selling*, page 260 Reprinted with permission of Tom Hopkins International, P.O. Box 1969, Scottsdale AZ 85252.

APPENDICES

Employment Agreement

This agreement is entered into at New York, NY on   , 19   , between XYZ INCORPORATED, a corporation of the State of New York, hereinafter referred to as "XYZ", and John Doe, hereinafter referred to as "EMPLOYEE."

With respect to any employment hereafter provided to EMPLOYEE by XYZ or any of its subsidiaries, and in consideration thereof, EMPLOYEE agrees with XYZ as follows:

FIRST: That all knowledge and information not already available to the public which EMPLOYEE may acquire respecting inventions, designs, methods, systems, improvements, trade secrets or other private or confidential matters of XYZ shall for all purposes be regarded by EMPLOYEE as strictly confidential and shall not, so long as it remains confidential, be directly or indirectly disclosed to any person without XYZ's prior written permission.

SECOND: EMPLOYEE further agrees on behalf of himself, heirs and representatives that he will promptly disclose and assign to XYZ or its nominee all inventions, patentable or unpatentable, conceived by him solely or jointly with others either (1) during his working hours, (2) at XYZ's expense or on XYZ's premises, or (3) at any time or place during the term of his employment, if the inventions relate to XYZ's business.

THIRD: He further agrees without compensation to assist in every proper way and to execute at any time whether or not he is still in the employ of XYZ all papers of his to secure for XYZ patent rights to such inventions in this and all foreign countries.

EMPLOYEE
Signature _____

WITNESS: _____

_____

Residence _____

City _____

State _____ Zip _____

APPENDIX **B**

Invention Submission Form (Employee)

To: Patent Department

Invention Disclosure

Title of Your Invention

Purpose of Your Invention

Brief Description of Your Invention (attach extra pages and drawings as needed)

Description of Closest Prior Art

Date of Conception of Your Invention

Date Disclosed to Others

Date of First Working Model

Date of First Sale or Offer to Sell

Plans for Publication or Commercialization

Signature _____

Date _____

Witnessed and Understood By:

_____   _____

Witness 1   Date

_____   _____

Witness 2   Date

APPENDIX C

Invention Submission
Form (Outsiders)

SUBMITTING IDEAS TO XYZ INCORPORATED

To avoid any possible future confusion between your ideas and those already acquired by XYZ and its subsidiaries from the efforts of their own employees and others, and to prevent any misunderstanding as to your rights and the obligations of XYZ, your submission can be considered only under the following conditions:

1. No suggestion will be considered unless it is submitted in writing. Material submitted will not be returned, so retain a duplicate.

2. No suggestion will be accepted by XYZ on the basis of a confidential relationship or under a guarantee that the idea shall be kept secret or on condition that XYZ shall agree to terms of compensation before they know the idea.

3. A patented idea, or one on which a patent application has been filed, will be considered only on the basis that the submitter will rely exclusively on the rights granted under the Patent Statutes. (In the case of a patentable idea, it is suggested that the submitter file a patent application before submitting an idea.)

4. Where an idea has not been patented, and no patent application is pending on it, payment, if any, for the use of the idea shall be in an amount solely within the discretion of XYZ or its subsidiary.

5. Neither XYZ nor its subsidiaries shall be obligated to give reasons for their decision or to reveal their past or present activities related to the submitted idea. Negotiating or offering to purchase an idea shall be without prejudice to XYZ or its subsidiaries and shall not be an admission of the novelty, priority, or originality of the idea.

TO XYZ INCORPORATED:

I have read the conditions outlined above and agree to them. In addition, I warrant that the idea submitted is my own and that I am free to offer it.

My idea is _____

_____

_____

_____

_____

_____

_____

_____

Signature _____

Date _____   Street Address _____

City & State _____ Zip _____

APPENDIX D

Secrecy and Patent Rights Agreement

The undersigned, an architect, engineer, consultant, contractor, sub-contractor, or supplier to THE XYZ COMPANY (hereinafter called "COMPANY"), or an employee of such architect, engineer, consultant, contractor, subcontractor, or supplier, in order to render services or supply material, machinery, or equipment in such capacity with respect to confidential and proprietary activities of COMPANY, in connection with proposed facilities for the manufacture of (____PRODUCT____) and in consideration of being given access by COMPANY to such confidential information and/or to premises, facilities, or documents embodying such confidential information, regarding manufacturing processes, equipment, compositions of matter, and use of said product, hereby agrees to the following terms and conditions.

1. All documents of any character including but not limited to drawings, designs, plans, specifications, requisitions, instructions, data, manuals, models, equipment, and the like, delivered to the undersigned by COMPANY or by an employee, agent, consultant, or contractor of COMPANY, or produced or developed by or for the undersigned pursuant to an Agreement or understanding with COMPANY, shall be and remain the property of COMPANY and shall be delivered by the undersigned together with all copies thereof to COMPANY promptly upon COMPANY's request provided, however, that the undersigned can retain one set of record copies.

2. It is agreed that any proprietary technical information disclosed shall not be used for the benefit of any third party or disclosed to any third party for a period of ten (10) years following disclosure without the consent of COMPANY, unless (1) such information is already public knowledge, (2) such information was in the possession of the undersigned prior to the disclosure by COMPANY, (3) such information after disclosure is published or otherwise becomes part of the public domain through no fault of the undersigned, or (4) such information was received after the time of disclosure hereunder from a third party who did not acquire it under an obligation of confidence directly or indirectly from COMPANY.

3. The disclosure of any information by COMPANY hereunder shall not confer any rights under any patents or patent application presently held or subsequently obtained by COMPANY in respect to inventions or technology relating to such disclosure.

4. COMPANY and the undersigned hereby agree to the following understanding with respect to inventions made by employees and/or agents of the undersigned while engaged in work done by the undersigned for COMPANY:

Any inventions, discoveries, or improvements of a process or apparatus nature so conceived and based on and directly related to information about the Process disclosed by COMPANY shall become the property of COMPANY.

The undersigned will do all lawful things necessary to permit COMPANY or its nominee or nominees to enjoy full ownership of such inventions, discoveries or improvements and agrees to use for the work only technical personnel who are bound to assign their inventions to the undersigned, to direct such personnel at COMPANY's request to assign such inventions and patent applications therefore to COMPANY, to use its best efforts to obtain the execution by such personnel of any necessary papers in connection with securing patents thereon, and, but without expense to the undersigned, to cooperate with COMPANY on the soliciting of such patent applications prepared by COMPANY and in the enforcement of patents issued thereon.

All other inventions, discoveries, or improvements conceived by the undersigned employees, which relate to general purpose or to specific processes or apparatus other than the Process disclosed by COMPANY, or which are not based on or directly related specifically to the Process disclosed by COMPANY, shall remain the property of the undersigned, and the undersigned shall be free to obtain patents thereon, but the undersigned agrees that COMPANY shall have a royalty-free license under any such patents to use such inventions in the Process.

The specification of any such patents of the undersigned shall describe the invention in a context not disclosing, among examples of uses of the invention, confidential information furnised to the undersigned by COMPANY.

AGREED

For _____

By _____

Title _____

Date _____

ACCEPTED FOR COMPANY

By _____

Title _____

Date _____

License Agreement

THIS AGREEMENT made and entered into on the _____
day of _____, 19_____, by and
between _____, having
its principal place of business at _____
_____, herein called "CORPORATION;"
and XYZ INCORPORATED, herein called XYZ,

WITNESSETH THAT:

In consideration of the mutual covenants herein contained and
intending to be legally bound hereby, the parties agree as follows:

1. DEFINITIONS

As used herein, the following terms shall have the meanings set forth
below:

A. *Inventions* means _____,
as disclosed and claimed in United States Patent Numbers _____
_____, and improvement inventions of XYZ.

B. *Patents* means United States and foreign Patents covering the
Inventions, patents to be issued pursuant thereto, and all divisions,
continuations, reissues, and extensions thereof.

C. *Technical Information* means the information and data in the
possession of XYZ that XYZ has the right to disclose relating to the
Inventions.

D. *Licensed Territory* means

E. *Licensed Field* means the

2. LICENSE

XYZ hereby grants to CORPORATION, to the extent of the *Licensed
Field* and *LicensedTerritory*:

A. A nonexclusive license to make, use, and sell products embodying
the *Inventions* under the *Patents*; and

B. A nonexclusive license to use the *Technical Information* in practicing
the *Inventions*.

3. INFORMATION-PATENTS

XYZ shall make available to CORPORATION for its use the *Technical Information* in its possession concerning the *Inventions* and shall promptly disclose to the other party on a nonconfidential basis all discoveries, developments, and patents pertinent thereto. CORPORATION hereby grants to XYZ, without additional consideration, a royalty-free, non-exclusive license with the right to sublicense all improvements of the *Inventions* which CORPORATION may have or acquire at any time during the term of this Agreement. As part of the consideration for this Agreement, XYZ agrees to disclose to CORPORATION XYZ's pending United States patent application disclosing the *Inventions* which are not otherwise available and which are held in confidence by the United States Patent Office. XYZ shall have the sole right to prosecute domestic and foreign patents on the *Inventions* and shall have the right to determine whether or not to file any patent application, to abandon the prosecution of any patent, or to discontinue the maintenance of any patent. CORPORATION shall reimburse XYZ for all foreign patent expenses within the *Licensed Field* which are incurred after the effective date of this Agreement. CORPORATION shall reimburse XYZ for the taxes and annuities in connection with the *Patents* in the *Licensed* Territory. CORPORATION shall notify XYZ in writing of its intent not to file a patent application, to abandon the prosecution of any application, or to discontinue maintenance of any patent in force on any improvements of the *Inventions* at least forty-five (45) days prior to the last day on which action is required to preserve such application from abandonment or to maintain such patent in force and upon written request shall promptly furnish to XYZ all papers pertaining to such application or patent, so that XYZ may have the option, at its own expense, to take the required action. XYZ shall, in those countries wherein CORPORATION is not working the *Inventions*, have the right to advertise for and grant licenses to third parties.

4. LITIGATION

CORPORATION shall notify XYZ of any suspected infringement of the *Patents* in the *Licensed Field* and *Licensed Territory*. The sole right to institute a suit for infringement rests with XYZ. CORPORATION agrees to cooperate with XYZ in all respects, to have any of CORPORATION's employees testify when requested by XYZ, and to make available any records, papers, information, specimens, and the like. Any recovery received pursuant to such suit shall be retained by XYZ.

5. ROYALTIES

A. CORPORATION shall pay to XYZ a royalty of _____ percent (%) of the net sales price of all products sold under the licenses granted under Article 2 hereof. In computing the net sales price, CORPORATION may deduct any sales commission paid to its sales representatives.

B. CORPORATION shall pay to XYZ, as a license fee, _____ Thousand Dollars ($_____), upon execution of this Agreement. Such license fee shall not be credited toward royalties under Paragraph A hereof.

6. MINIMUM ROYALTIES

CORPORATION shall pay to XYZ royalties as stated in Paragraph A of Article 5, but in no event shall annual royalties from sales in the *Licensed Territory* be less than the following minimum royalties in each of the calendar years indicated.

Calendar Year	Minimum Royalties
1985	_____
1986	_____
1987	_____
1988	_____
1989 and each calendar year thereafter during the term of this Agreement	_____

7. PAYMENTS

A. Not later than the last day of each February, May, August, and November CORPORATION shall furnish to XYZ a written statement in such detail as XYZ may reasonably require of all amounts accrued hereunder during the quarterly periods ending the last day of the preceding December, March, June, and September, respectively, and shall pay to XYZ all amounts due to XYZ. In the event that the amounts paid and accrued as of December 31 in any calendar year do not equal the minimum royalty specified in Article 6 for such year, CORPORATION shall pay to XYZ on the last day of February next following such December 31, the amount required to satisfy its minimum royalty obligation for such year. Such amounts are due at the date the reports are due. If no amount is accrued during any quarter, a written statement to that effect shall be furnished.

B. All payments shall be in U.S. dollars unless XYZ notifies COR-PORATION of its intent to receive such payments in other currencies. For the purpose of this Agreement, the rate of exchange to and from U.S. dollars for any currency of other countries shall be mutually agreed upon, and if the parties fail to agree, at the par value established by the International Monetary Fund, as of the last business day of the calendar quarter in which CORPORATION receives such payment.

C. Royalty payments provided for in this Agreement shall, when overdue, bear interest at the prime rate plus one percent (1%) per annum as established by Manufacturers Hanover Trust in New York City from the date due until payment is received by XYZ.

D. If this Agreement is for any reason terminated before all of the payments herein provided for have been made, CORPORATION shall immediately pay to XYZ any remaining unpaid balance even though the due date as above provided has not been reached.

8. RECORDS

CORPORATION shall keep accurate records of all operations affecting payments hereunder, and shall permit XYZ or its duly authorized agent to inspect all such records and to make copies of or extracts from such records during regular business hours throughout the term of this Agreement and for a reasonable period of not less than three (3) years thereafter.

9. REPRESENTATIONS

A. XYZ represents that it has the right to grant all of the rights granted herein.

B. Nothing in this Agreement shall be deemed to be a representation or warranty by XYZ of the validity of any of the *Patents* or the accuracy, or usefulness for any purpose, of any *Technical Information*, techniques, or practices at any time made available by XYZ. XYZ shall have no liability whatsoever to CORPORATION or any other person for or on account of any injury, loss, or damage, of any kind or nature, sustained by, or any damage assessed or asserted against, or any other liability incurred by or imposed upon CORPORATION or any other person, arising out of or in connection with or resulting from (i) the production, use, or sale of any apparatus or product or the practice of the *Inventions*, (ii) the use of any *Technical Information*, techniques, or practices disclosed by XYZ, or (iii) any advertising or other promotional activities with respect to any of the foregoing.

10. TERMINATION

A. This Agreement shall end upon the expiration of the last to expire of the *Patents* included herein, or upon the abandonment of the last to be abandoned of any patent applications included herein, whichever is later, unless this Agreement is sooner terminated.

B. CORPORATION may terminate this Agreement at the end of any calendar year upon sixty (60) days written notice in advance to XYZ or at any time after the expiration of any patent included herein or a final adjudication of invalidity of any patent included herein by written notice in advance to XYZ.

C. If either party shall be in default of any obligation hereunder, or shall be adjudged bankrupt, or become insolvent, or make an assignment for the benefit of creditors, or be placed in the hands of a receiver or a trustee in bankruptcy, the other party may terminate this Agreement by giving sixty (60) days notice by Registered Mail to the party at fault, specifying the basis for termination. If within sixty (60) days after the receipt of such notice, the party at fault shall remedy the condition forming the basis for termination, such notice shall cease to be operative and this Agreement shall continue in full force.

D. If CORPORATION contests the validity of any of the *Patents*, said contest shall be considered a notice of termination of the Agreement effective as of the date (1) of initiation of said contest, or (2) validity is raised as a defense by CORPORATION in any action brought by XYZ to enforce any obligation under the Agreement; and provided further that CORPORATION shall, if the patent validity is upheld, reimburse XYZ for its costs, legal fees, and other expenses relating to said validity contest.

11. MARKETING

CORPORATION shall place in a conspicuous location on any product made or sold under any patent coming within this Agreement a patent notice in accordance with the laws concerning the marketing of patented articles.

12. PUBLICITY

In publicizing anything made or sold under this Agreement, COR-PORATION shall not use the name XYZ or otherwise refer to any organization related to XYZ, except with the written approval of XYZ.

13. NONASSIGNABILITY

CORPORATION shall not assign any right hereunder without the written consent of XYZ.

14. REFORM

The parties agree that if any part, term or provision of this Agreement shall be found illegal or in conflict with any State or Federal law, the validity of the remaining provisions shall not be effected thereby.

15.   WAIVER AND ALTERATION

A. The waiver of a breach hereunder may be effected only by a writing signed by the waiving party and shall not constitute a waiver of any other breach.

B. A provision of this Agreement may be altered only by a writing signed by both parties.

16.   IMPLEMENTATION

Each party shall execute any instruments reasonably believed by the other party to be necessary to implement the provisions of this Agreement.

17.   CONSTRUCTION

This Agreement shall be construed in accordance with the laws of the State of New York of the United States of America and in the English language.

18.   EXPORTATION OF TECHNICAL INFORMATION

CORPORATION agrees to comply with the laws and rules of the United States Government regarding prohibition of exportation of reports and technical information furnished to CORPORATION either directly or indirectly by XYZ.

19.   ENTIRE UNDERSTANDING

This Agreement represents the entire understanding between the parties, and supersedes all other agreements, express or implied, between the parties concerning the *Inventions*.

20. ADDRESSES

For the purpose of all written communications between the parties, their addresses shall be:

_____

_____

_____

XYZ INCORPORATED

_____

_____

_____

or any other addresses of which either party shall notify the other party in writing.

IN WITNESS WHEREOF the parties have caused this Agreement to be executed by their duly authorized officers on the respective dates and at the respective places hereinafter set forth.

ATTEST:

By: _____   By: _____

                               Title: _____

Signed at: _____   Date: _____

ATTEST   XYZ INCORPORATED

By: _____   By: _____

Signed at: _____

                               Date: _____

F

Joint Venture
(Considerations)

PROPOSED PARTNER

1. Secrecy areement—for exchange of technical and business sensitive information (five years)—prepared by Patent Department.

   Problems—Our Company cannot be precluded from European or Japanese markets if a deal is not consumated.

2. Type of organization
   Corporation?
   Partnership?

3. Scope
   A. Product Definition
      1. Marketing
         Distributorship agreements and/or direct sales
      2. Packaging–Labeling
      3. Manufacturing license—at what stage?
   B. Other products, improvements and further developments
         Option to obtain further patents at what royalty rates
   C. Grant Back and U.S. Distribution of Improvements

4. Location—Corporation of what country?
   Tax advantages
   Management advantages
   Location of facilities
   Member of Common Market
   Government support, etc.

5. Territory
   World
   Asia
   Europe
   United States

6. Financial
   Capitalization
   Dividend policy
   Reporting system
   Loans—Loan agreement from Parents

7. Exclusivity
  Territory
  Diligence
8. Term
  Infinite or limited?
  Termination provisions
  Confidentiality of information generated by J.V.
  Arbitration
9. Management
  A. Composition of the board
    Equal representation
    Equal + 1 (independent general manager)
    Restrictions on management
      Majority of board approval for:
      Loans greater than _____ dollars
      New products
      Leases
      Contracts exceeding _____ dollars
  B. Selection of Officers
    CEO
      From parent?
      From outside?
      Independence?
    Selection of controller?
    Selection of other officers?
  C. Management and technical assistance contracts
10. Manpower
  Borrowed from Parent?
  Hired from Parent?
  Hired from outside?
  What's needed initially?
11. Functional responsibilities
  Administrative
  Manufacturing/packaging/assembly
  Marketing
  Distributorship agreements—forms
  Direct sales

Use of trademarks—license
  Parents
  J. V.'s
Research and development
  Scope
  Where?
  Facilities—overhead
Patents, foreign filing, joint
General legal—provided by whom?

12. Licensing
    From Parents
    Grant back
    License to others
    What terms and royalties

13. Transfer prices
    Between J.V. and Parents
    Proper focus
    Terms of sales

14. Additional products for future
    Acquisitions?
    Self developed?
    Parents?

15. Relative value/importance of contributions of each party
    Technology
      Know how
      Parents?
    Marketing
      Distribution
      Organization
      Trademarks
    Financial contributions
      Profit Share
      Equal

Resume—Director of Research

Gerald Garner
41 Safe Harbor Ct.
Buffalo, New York 14100

1980–Present THE CARBON COMPANY

Director of corporate wide effort to market energy related products
and services. Strategy includes identification of target industries,
new product needs, new process development, energy availability,
cost analysis, and identification of firms as possible acquisition
candidates or competitors. An energy efficient ladle preheater is
presently being introduced to the market.

1974–1980 THE CARBON COMPANY

Associate Director of Research and Development, directed the
technical activities of 125 scientists and supporting staff. Over 100
patents were issued to the Division. Organized ceramic, chemistry,
engineering, analytical departments.

Initiated numerous R&D projects and brought them to commercialization including:

- Ceramic armor, with a process for manufacture which resulted in
  $10 million annual sales.
- A high performance aromatic polymer, which formed the base for
  a major new Division.
- Solid state doping agents for electronic silicon, annual sales over
  $1 million.
- A flame resistant phenolic fiber, commercialized through an
  international joint venture.
- Reduced raw material cost by 50% for heating elements by
  developing a new form of silicon carbide.

Developed working relationships with university, government, and
independent laboratories and professional societies to augment our
R&D programs.

1968–1974   AMERICAN TEXTILE, INC.

Technical Director for the subsidiary, American Textile, Inc. Brought research product to the marketplace by interfacing with customers, regulatory agencies, marketing and distribution systems, internal and subcontract manufacturing, and general technical business management. Invented a new process for the curing of textile fibers which reduced the cost of manufacture by two-thirds.

1963–1968

As Research Manager, initiated exploratory projects which led to new business in ceramics, inorganic and polymer chemistry, rocket nozzles, nuclear fuels and poison pellets, pollution control processes, electronic materials, textile fibers, and fire safe materials.

1960–1963

As Manager of the Physics Department, conducted studies of the properties of silicon carbide leading to improved heating elements, thermistors, and other electrical devices.

1958–1960

As Senior Research Associate, studied the growth and properties of single crystals of silicon carbide.

1954–1958   OLSON CHEMICAL CORP.

As Group Leader and Research Associate, developed several alternative manufacturing processes for diborane. One process produced the nation's needs for this precursor of a rocket fuel for several years.

EDUCATION

Advanced Management Program, Harvard Business School
Ph.D, Physical Chemistry, University of Michigan
B.A., Physics and Chemistry, Cornell University

APPENDIX H

Resume—Director of Development

Bruce F. Michel
770 Bishop Road
Boston, MA

Home (617) 555-1212
Business (617) 555-0000

SUMMARY

Experienced in market assessment, developing and packaging ventures, customer and competitor analysis, risk evaluation, financial assessment, and presentation of opportunities to the Board of Directors.

EXPERIENCE

Arden Ceramic Materials, Co., 1980
Vice President, New Business Development

Active participant in the formation of a new company to manufacture ceramic materials, including silicon nitride and boron nitride. Elected to Board of Directors, May 1981. Annual sales exceeded $3 million within 2 years.

The Connecticut Corporation, 1977–1980
Director of New Business Development

Directed the integration of corporate technologies and initiated the business development function at the Brown Development Center.

Developed opportunities and transferred ventures to operating divisions. Ventures included:

- A technology for making metal strip.
- A ceramic fiber dielectric separator for lithium–sulfur batteries.
- A government development program, starting with bench R&D to prototype plant for a new separator—$1.4 million over 2 years.
- An aluminum reduction cell.

The Vincent Company (Acquired bt the Connecticut Corporation), 1961–1977
Manager of Operations

Directed the start-up of a widget business from the laboratory to a 35 million pound/year chemical plant.

Negotiated exclusive U.S. licenses for pollution control equipment from major German and Japanese companies. This became the nucleus for a waste water treatment business.

Spearheaded a successful effort to transfer a university invention to a commercial product. This was the start of an analytical services business generating $3 million in annual sales.

Developed a divestment strategy and directed the sale of a pollution business.

Managed the Ceramic Department and directed a new products operation which generated annual sales of over $2 million.

Union Metals Company, 1957–1961

Worked in the area of metal chemicals development.

EDUCATION

M.B.A., University of Rochester
M.S., Columbia University (Chemistry)
B.S., Columbia University (Chemistry)

PATENTS AND PUBLICATIONS

Several published articles in the areas of high temperature materials and six patents.

PROFESSIONAL SOCIETIES
American Chemical Society
American Ceramic Society
American Association for the Advancement of Science

APPENDIX I

Consultant Agreement

This AGREEMENT is executed as of _____, by _____ (hereinafter called CONSULTANT) and XYZ INC., a corporation of New York (hereinafter called XYZ).

WITNESSETH:

WHEREAS, XYZ desire to obtain the CONSULTANT's services to aid XYZ in the performance of experimental and research investigations in the field of _____; and

WHEREAS, the CONSULTANT desires to aid XYZ by performing such services (hereinafter called the "WORK");

NOW, THEREFORE, in consideration of the mutual covenants and promises herein contained, the parties hereto agree as follows:

ARTICLE I—STATEMENT OF WORK

The CONSULTANT shall perform WORK on experimental and research investigation in the field of _____, and such other WORK as mutually agreed upon. The CONSULTANT agrees to perform such services as may be requested by XYZ hereunder to the best of his ability and during regular business hours. This agreement is made with the understanding that CONSULTANT is an independent contractor and not an employee of XYZ. CONSULTANT, an employee of a University, warrants that his services hereunder are not in confilict with nor prohibited by his University employment.

ARTICLE II—PAYMENT

For performing the WORK, CONSULTANT shall be paid _____ dollars ($) per day for the days he spends in actual WORK. XYZ agrees to utilize and CONSULTANT agrees to perform at least three (3) days of actual WORK per year or a greater number of days as mutually agreed on but not to exceed a total of ten (10) days unless additional services are authorized in writing by an officer of XYZ. Such payments shall constitute full payment to the CONSULTANT for all services performed hereunder, except that XYZ shall, upon submission of an itemized statement covering the same by the CONSULTANT, reimburse the CONSULTANT for all actual expenses approved in advance in writing by XYZ and paid by the CONSULTANT for travel, telephone, and telegraph in the performance of services requested by XYZ hereunder.

ARTICLE III—TERM

The Term of this Agreement shall be one year from its effective date of _____, unless sooner terminated as provided for herein, or extended by mutual agreement in writing. Any other modification of the terms of this Agreement shall not be binding upon XYZ unless accomplished by a formal written supplement to this Agreement.

ARTICLE IV—REPORTS

The CONSULTANT shall furnish reports concerning activities of the CONSULTANT under this Agreement to XYZ from time to time, in such form as may be reasonably required by XYZ.

ARTICLE V—PERSONAL SERVICES— ASSIGNMENT

A. The WORK or services provided for herein shall be performed personally by _____ or such other person approved in advance by XYZ, and no other person shall be engaged upon such WORK or services by the CONSULTANT; provided, however, that this paragraph shall not apply to secretarial, clerical, and similar incidental services needed by CONSULTANT to assist him in performance of this contract and provided at his own expense unless such expense is approved in writing in advance by XYZ.

B. Neither this contract, nor any interest therein or claim thereunder, shall be assigned or transferred by CONSULTANT to any party or parties without the written authorization of XYZ.

ARTICLE VI—PATENT, COPYRIGHT, AND DATA RIGHTS

A. CONSULTANT shall promptly disclose in writing to XYZ all ideas, inventions, improvements, and developments relative to the field of WORK originated by him in connection with his WORK under this Agreement. CONSULTANT will upon request and without additional compensation execute all papers necessary to transfer to XYZ, free of encumberance or restrictions, all inventions, discoveries, and improvements whether patentable or not, conceived or originated by CONSULTANT or with others relating to the PRODUCTS. All transfers aforesaid shall include the patent rights in this and all foreign countries.

B. In respect to all copyrightable material, first produced or composed by CONSULTANT in connection with the performance of his WORK under this Agreement, CONSULTANT hereby conveys to XYZ the sole and exclusive right to transfer or obtain for itself or its designee any and all rights to such copyrightable material as it shall deem appropriate in its sole and exclusive discretion, agrees to and hereby assent to all such transfers, and agrees to assist in the registration and/or transfer of all such copyrights by XYZ or its designee in any manner as aforesaid.

ARTICLE VII—TERMINATION

This Agreement may be terminated by either party by giving to the other party thirty (30) days written notice.

Upon such termination, CONSULTANT shall be reimbursed for all his approved expenses incurred prior to such termination. Termination shall not affect CONSULTANT's obligations under Articles IV, VI, and VIII.

ARTICLE VIII—SECRECY

CONSULTANT agrees that for a period of five (5) years after conclusion of this Agreement he will keep confidential any confidential information of XYZ obtained by him during the period in which this Agreement is in force and to refrain from using, publishing, or revealing such information acquired by him in the course of this WORK without the written consent of XYZ, excepting only information (1) that was known to CONSULTANT prior to its disclosure by XYZ, (2) that becomes known to the public without fault of CONSULTANT, (3) that is disclosed to CONSULTANT by a third party in good faith, or (4) that is specifically released from confidential status by XYZ.

IN WITNESS WHEREOF, the parties hereto have executed this Agreement as of the date and year first above written.

By: _____

Date: _____

XYZ INC.

Title: _____

_____

Date _____

Management by Objectives Worksheet

PERFORMANCE OBJECTIVES AND GOALS

Name _____

Title _____

Department _____

Division _____

INSTRUCTIONS

1. The purpose of this form is to: (a) have the employee participate in the establishment of his performance objectives and development goals, (b) continuously monitor and review goals and performance, and (c) accelerate professional growth by pinpointing strengths as well as problem areas.

2. List your major objectives and development goals on Part A. These should be realistic and stated in specific rather than general terms. Indicate how results are to be measured.

3. After completing Part A, you should submit this form to your supervisor for discussion and approval. When agreement is reached, copies of the form should be signed and retained by you and your supervisor.

4. Periodically, you and your supervisor should discuss progress on each item.

5. By the end of the year, you should prepare a brief report on your progress using Part B. All objectives and plans should be reviewed. To what extend have you accomplished—or failed to accomplish—your goals. Be specific.

Management Development Form—Employee Section

SELF-DEVELOPMENT PLAN To be complete and
reviewed with your manager)

Name _____ Title _____ MANAGER

_____

Major accomplishments during past year

_____

Your strengths

_____

Self-development actions completed during past year

_____

Development needs during next year—plan for next year

_____

Career interests

Next assignment _____ Longer range _____

_____

Additional comments

PART A

Major Objectives

List major objectives. These should be realistic and specific. Indicate how
results are to be measure.

Development Goals

Objectives and Personal Goals Agreed to

Employee _____   Date _____
Supervisor _____   Date _____

Part B

Performance Report

State results achieved versus objectives. Explain variances. Utilize quantitative measures whenever possible.

Development Goals

Career Objectives

What positions are you interested in or qualified for?

Now:
Within 2–5 Years:

Report Completed and Discussed with Supervisor

Employee _____   Date _____
Supervisor _____   Date _____

Questions Frequently Asked During a Job Interview

1. What are your short range and long range objectives?
2. What do you expect to be doing in 5 years?
3. Why are you looking for a new position?
4. Why did you leave your last position?
5. What do you like and dislike about your present job?
6. What do you like and dislike about your present employer?
7. What are your major strengths and weaknesses? Can you give me at least three of each?
8. Would you tell me about your major accomplishments at work and in general?
9. How long would it take you to make a significant contribution to our company?
10. If you could go back to the beginning of your career, what changes would you make?
11. Tell me about some of your mistakes.
12. In your opinion, what are the attributes of a good boss?
13. Why do you think that you have management potential?
14. What was the best book that you read during the past year?
15. What types of books do you enjoy?
16. What are your hobbies?
17. What do you look for in subordinates and why?
18. What were you favorite subjects in college? What subjects did you dislike?
19. Tell me about a technical problem that you have solved and how you went about it.
20. How many invention disclosures have you submitted in the past 3 years?
21. What would you do to motivate a group of engineers?
22. Why are you interested in this position and in our company?
23. Tell me about your outside interests.

Questions to Ask a Prospective Supervisor

1. What are your company's major strengths and weaknesses?
2. How would you describe the company's short range and long range objectives?
3. Could you be more specific with respect to the company's goals?
4. What type of people are at the top—primarily financial, marketing, or generalists?
5. Was top management promoted from within or hired from outside?
6. How does the company compare with its major competition?
7. What is the company's status in the industry?
8. How much emphasis does the company place on new products and technology?
9. Would you describe your organization's mission and how it fits into the overall corporate plan?
10. What are your organization's 1 and 5 year goals?
11. Would you tell me about some of your organizations major accomplishments?
12. What are your department's major strengths and weaknesses?
13. Has your organization made a meaningful contribution to sales? How?
14. Has your organization made a major contribution to manufacturing?
15. How effective has your group been in cost reduction and quality improvement?
16. Why are you going outside to fill this position?
17. How would you describe your present superior?
18. Would you tell me about your management philosophy?
19. How would you define the traits of a good manager?
20. How would you define your ideal candidate for the present position?
21. Can you tell me about the job description for this position?
22. Assuming that I did an outstanding job for you, are there opportunities for promotion?

23. Tell me about your company's approach to human resource development.

24. Why do you think that your company is a good place to work?

Index

275